STRUCTURAL CHARACTERISTICS
OF MATERIALS

ELSEVIER MATERIALS SCIENCE SERIES

Advisory Editors

LESLIE HOLLIDAY,

Visiting Professor, Brunel University,
Acton, London, Great Britain

A. KELLY Sc.D.,

Deputy Director, National Physical Laboratory,
Teddington, Middlesex, Great Britain

STRUCTURAL CHARACTERISTICS OF MATERIALS

Edited by

H. M. FINNISTON, F.R.S.

Deputy Chairman, British Steel Corporation

ELSEVIER PUBLISHING COMPANY LIMITED

AMSTERDAM—LONDON—NEW YORK

1971

ELSEVIER PUBLISHING COMPANY LIMITED
BARKING, ESSEX, ENGLAND

ELSEVIER PUBLISHING COMPANY
335 JAN VAN GALENSTRAAT, P.O. BOX 211, AMSTERDAM
THE NETHERLANDS

AMERICAN ELSEVIER PUBLISHING COMPANY INC.
52 VANDERBILT AVENUE, NEW YORK, N.Y. 10017

0 444 20045 2

LIBRARY OF CONGRESS CATALOG CARD NUMBER 79–80492

With 4 Tables and 215 Illustrations

Printed in Great Britain by Galliard Limited, Great Yarmouth, Norfolk, England.

LIST OF CONTRIBUTORS

A. G. CROCKER
Department of Physics, University of Surrey, Guildford, Surrey

G. J. DAVIES
Department of Metallurgy, University of Cambridge

D. HULL
Department of Metallurgy and Materials Science, University of Liverpool

H. LIPSON
Department of Physics, The University of Manchester Institute of Science and Technology

S. G. LIPSON
Institute of Technology, Haifa, Israel

R. B. NICHOLSON
Department of Metallurgy, University of Manchester

WALTER S. OWEN
Dean, The Technological Institute, Northwestern University, Evanston, Illinois

E. ROBERTS
Department of Metallurgy, University of Liverpool

FREDERICK J. SCHOEN
Gulf General Atomic, San Diego, California

J. WILLIAMS
Atomic Energy Research Establishment, Harwell, Berks

B. T. M. WILLIS
Atomic Energy Research Establishment, Harwell, Berks

CONTENTS

Chapter 2

DEFECT STRUCTURES—*A. G. CROCKER*

Chapter 3

MECHANICAL TWINNING—*D. HULL and E. ROBERTS*

CONTENTS

CHAPTER 1

CRYSTALLOGRAPHY OF MATERIALS

H. LIPSON AND S. G. LIPSON

1. INTRODUCTION

For most ordinary uses of materials we need to know only macroscopic properties such as the elastic moduli and the limits over which they can be applied. For a complete understanding of the properties of matter a knowledge of the underlying atomic arrangements is necessary. Until 1912 the possibility of obtaining this information was remote; but the discovery in that year by Laue, Friedrich and Knipping of the diffraction of X-rays by crystals, and its subsequent exploitation by W. L. Bragg,[1] opened up a new field of study, the implications of which are still not completely worked out.

Basically, the method showed that almost all solid matter is crystalline. X-ray diffraction patterns of perfect crystals could yield complete information about the arrangements of atoms within them. Studies of such crystals have abounded and, although there is no general solution to the problem of deriving a crystal structure from its X-ray diffraction pattern, so many new techniques and devices—including digital computers—are available, that it is almost true to say that any problem of reasonable complexity can be solved if enough effort, in manpower and money, can be put into it.

Such studies have reaped rich rewards in almost all the branches of science to which they have been applied. Chemistry, for example, was revolutionised by Bragg's discovery that molecules of NaCl did not exist in rock salt.[2] We have probably reached the end of an epoch of basically important discoveries in the single-crystal field, and the methods must now be extended to the more difficult problems of imperfectly crystalline matter. The most useful metallic materials, for example, have somewhat imperfect structures, and polymers are very imperfect indeed. Early attempts to investigate such structures, although they produced some important ideas, were not very informative; but with our present knowledge of the principles of diffraction the subject can be put on a much more systematic basis.

This basis must nevertheless contain the principles of the diffraction of X-rays by perfect single crystals, and it is the purpose of this chapter to set out these principles, and some of the results from them, in order to lay the foundations for the work described in the later chapters of this book.

2. CRYSTAL SYMMETRY

2.1. Periodicity

The study of the symmetry of crystals is a fascinating subject in its own right, and a series of books could be devoted to it. The basic principles were fully understood before there was any hope of testing them in practice, and they helped greatly in laying the foundations of the subject of crystal-structure determination when, after the first few years of groping, it was

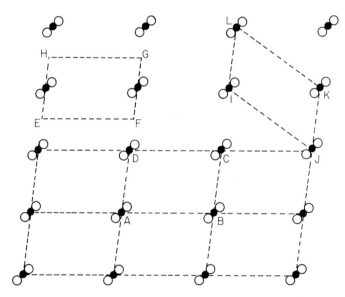

Fig. 1. Representation of a two-dimensional structure showing possible choices of unit cell (*ABCD*, *EFGH*, *IJKL*). (From Bunn, *Chemical Crystallography*, Clarendon Press, 1961, p. 14.)

realised that these principles existed. We cannot do more here than to provide an outline of the subject; but we shall try to do this in a systematic, rather than a descriptive, manner. It is easy to enumerate the crystal systems, crystal classes and so on, but it is more important to show how they arise.

We must accept certain experimental facts. The first is that a single crystal has a three-dimensionally periodic structure; in other words, a crystal consists of a group of atoms—or even one atom—repeated in exactly parallel orientation at exactly regularly-spaced positions. This is illustrated in two dimensions in Fig. 1.

From this figure two important concepts arise. The first is that of the lattice. If we take any representative point within the basic group, we see

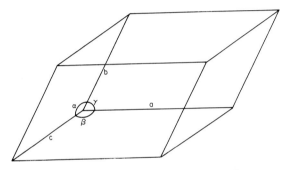

Fig. 2.   Axes and angles of a unit cell.

that this is repeated at regular intervals, and so the complete set—called the *space lattice*—forms a useful basis for consideration of the crystal structure, even if we do not know the positions of the atoms themselves. A space lattice is defined as the point intersections of three sets of regularly-spaced planes. It is an abstraction of considerable theoretical importance; but it should not be used as a substitute for 'structure'. Unfortunately, the word is so evocative that it is often so used, particularly by metallurgists; this use is equivalent to replacing a diffraction grating by its spacing. The usage is not only wrong; it can also be misleading.

The second important concept is that of the *unit cell*. This is most simply described as the unit of the space lattice and is always a parallelepiped. The unit cell of the two-dimensional structure shown in Fig. 1 is illustrated in the diagram. The three-dimensional counterpart (Fig. 2) is described either by three vectors **a**, **b**, **c**, or by the moduli of these vectors (a, b, c) and the angles (α, β, γ) between them.

2.2. Symmetry

All these results were deducible—and were deduced—before there was any direct demonstration of crystalline periodicity; they provide the only obvious basis for the observation that crystals grow with plane faces and that crystals of a particular material, grown under similar conditions, always have the same interfacial angles whatever their shape.

There is another experimental fact that must also be taken into account: most crystals are symmetric. Drawings of some crystals are shown in Fig. 3,

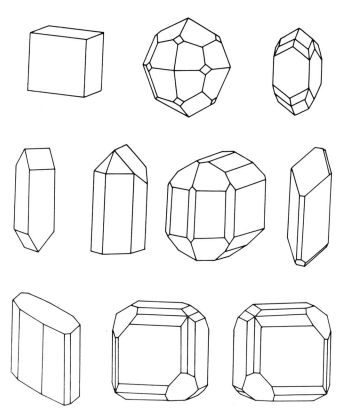

Fig. 3.   Some crystalline shapes. (From Bunn, *Crystals*, Academic Press, New York, 1964, p. 186.)

and here the symmetries are shown by the shapes; if we rotate the cube shown in Fig. 3 by 120 degrees about its diagonal it would look exactly similar. In practice, however, crystals do not usually grow so symmetrically, and one has to delve more deeply to recognise their symmetry; the arrangements of *face normals* are more fundamental.

What is the significance of this symmetry? It can arise from two causes: first, the unit group of atoms may itself be symmetric and some or all of its symmetries may be preserved as it unites with other groups to form a crystal; secondly, the group may be unsymmetric, but it may unite with a few others to form a symmetric arrangement. The number of possible symmetries is quite small, in spite of the fact that an infinite number of them can be envisaged.

The reason for this limitation is that the only symmetries possible are those that can conform with parallel repetition. Suppose, for example, that a pattern has an *axis of symmetry*; that is, if we rotate it through a submultiple of 2π its appearance will not change (Fig. 4). It is easily

Fig. 4(*a*)

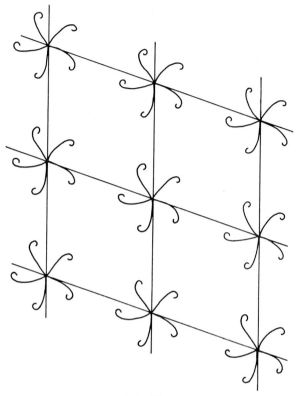

Fig. 4(*b*)

Fig. 4. (*a*) Pattern with two-fold axis of symmetry; (*b*) Pattern with five-fold axis of symmetry, showing that the periodically-repeated pattern does not have this symmetry.

possible to see that, if the rotation is $2\pi/2$ (Fig. 4(*a*)), the complete periodically-repeated pattern also has this symmetry, but if it is $2\pi/5$ (Fig. 4(*b*)) it has not. In fact, the only possible rotations consistent with regular repetition are $2\pi/1$, $2\pi/2$, $2\pi/3$, $2\pi/4$ and $2\pi/6$. These symmetries are called 1-fold, 2-fold, etc., and are denoted by the symbols 1, 2, 3, 4 and 6. The first symbol obviously indicates that the body possesses no symmetry.

Other types of symmetry are also possible. Some of the crystals shown in Fig. 3 have faces that exist in parallel pairs. They are obviously not connected by rotation and a new symmetry operation—*inversion*—has to be adduced. This is indicated by the symbol $\bar{1}$, and can occur in conjunction with any of the rotation axes. For example, the tetrahedron shown in Fig. 5 has what is called a fourfold inversion axis, $\bar{4}$.

All the symmetries possible in crystals number ten, denoted by the symbols 1, 2, 3, 4, 6, $\bar{1}$, $\bar{2}$, $\bar{3}$, $\bar{4}$ and $\bar{6}$.

It may be thought that one very obvious symmetry has been omitted— the mirror plane; patterns in which one half is the mirror image of the

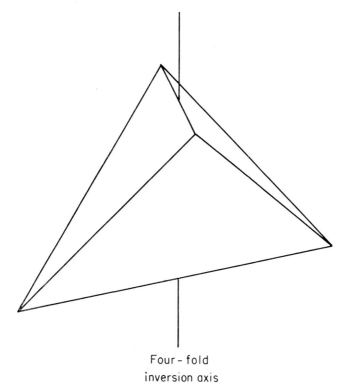

Four-fold
inversion axis

Fig. 5.   Tetrahedron showing four-fold inversion axis.

other half—the human face, for example—spring readily to the mind. But it can be seen that $\bar{2}$ *is* this symmetry and since it is of especial importance it is given the symbol *m* rather than $\bar{2}$.

2.3.   External symmetry of crystals

It does not follow that the ten symmetry elements lead to ten different types of symmetry in crystals; there are more than ten because they can exist in combinations. For example, 2 can exist with *m* if they are mutually perpendicular; this arrangement has the symbol $2/m$. Altogether 32 combinations of symmetry elements can be shown to exist; they are called *point groups*.

It is beyond the scope of this chapter to try to derive these 32 point groups, but some of the important ones are worth describing. For example, there is the point group $m3m$: the first symbol states that there is a mirror plane perpendicular to an axis, and the final symbol states that there is another mirror plane at some angle other than $\pi/2$ to the first; the symbol 3 states that there is a threefold axis in some other direction. It turns out that the only system of axes consistent with these statements is cubic, because a cube has mirror planes perpendicular to its edges and to its face diagonals, and threefold axes along its body diagonals.

Another important example is the point group $6/mmm$. The first symbol indicates a sixfold axis, the second indicates a perpendicular mirror plane, and the third and fourth indicate two sets of mirror planes containing the sixfold axis at an angle of $\pi/12$ to each other.

These statements may appear rather mystifying, indicating that there is a great deal of arbitrary nomenclature to learn. In fact, a full treatment[3] shows that the whole system is quite straightforward, and if one learns the system for the simpler point groups, the more complicated ones follow.

2.4. Crystal systems

Crystals can be assigned to the 32 crystal classes by experimental methods that we shall not describe here. It is convenient, however, to group the classes into seven crystal systems, each system being characterised by particular properties of the shape of the unit cell. For example, if a crystal has symmetry 1 or $\bar{1}$, the unit cell is of arbitrary shape (triclinic) and there are no relations between the unit-cell parameters, a, b, c, α, β, γ. If, however, the symmetry is 2, m or $2/m$ one axis must be perpendicular to the other two; if we accept the convention that the unique axis is b, we can state that $\alpha = \gamma = \pi/2$. Such crystals belong to the monoclinic system. In the orthorhombic system, for which the unit cell is a rectangular parallelepiped, the possible symmetries are mm, 222, and mmm. (The combination 22 is not possible because two perpendicular twofold axes beget a third.)

The other systems are tetragonal (perpendicular axes with $a = b$), trigonal and hexagonal, for which $a = b$, $\alpha = \beta = \pi/2$, and $\gamma = 120°$, and cubic. The last is the most interesting; its essential symmetries are the threefold axes along the cube diagonals; it does not necessarily have four-fold symmetry as one might expect.

2.5. Internal symmetry of crystals

So far, we have dealt entirely with properties that can be deduced from the measurements of the interfacial angles of crystals. We have seen that the symmetry of a crystal is decided by the symmetry of the atomic arrangements of which it is composed, and it might therefore seem that, conversely, knowledge of the crystal class would give all the necessary information about the symmetry of the internal atomic arrangement.

This is not necessarily true. We can say that a finite body has symmetry if the continued operation of a symmetry element ultimately brings the body back into its original state exactly; n operations of n-fold rotation obviously do so. But for an infinite body with a periodic structure other symmetry operations are possible; the body will still appear similar if it is, for example, rotated by π and shifted by half a lattice translation. Two such successive operations will produce the same arrangement shifted by a complete lattice translation. Such an operation is called a *screw axis*, and is denoted by the symbol 2_1. Similar considerations apply to *glide planes*, which involve reflexion and translation parallel to the plane o reflexion; the symbol m is then replaced by symbols such as a, b, c or n, depending upon the direction of the glide.

2.6. Space groups

With these symmetries, the number of possible combinations increases considerably. Each combination is called a *space group*,[3] and the number possible is 230. In the triclinic system there are only two space groups, corresponding to 1 and $\bar{1}$, but in the monoclinic system there are many more, because some or all of the rotation axes and mirror planes can be replaced by screw axes and glide planes. For example, crystals belonging to the point group 222 may have any of the four symmetries 222, $2_1 22$, $22_1 2_1$ or $2_1 2_1 2_1$. The combinations of mirror planes cannot be dealt with so simply.

Another complication is possible: symmetries may be present in combination as well as in substitution. For example, a crystal belonging to the point group 2 may have the internal symmetry 2_1, or it may have alternating 2 and 2_1 axes. It is easy to show that the addition of such symmetry elements leads to what are called *non-primitive* lattices.

2.7. Lattices

In Section 2.1 it was stated that the lattice of a crystal was a collection of regularly-spaced points in three dimensions; associated with each point is a collection of atoms, all similar to each other and in parallel orientation. *All* crystals can be described in this way. But it is convenient to describe some crystals in other ways. For example, it can be shown that if a crystal has the combination 2 and 2_1, the lattice points are no longer directly related by the lattice translations; another lattice point is produced in the centre of one of the faces of the unit cell, in the sense that the arrangement of atoms around this point is exactly the same as that around the origin.

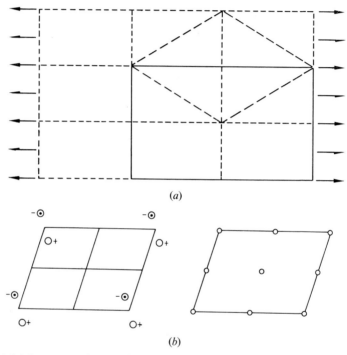

(a)

(b)

Fig. 6. (*a*) Symmetry elements in a space group with two-fold rotation and screw axes. The full line shows a rectangular centred unit cell; the broken line shows a non-centred cell which is not rectangular. (*b*) Representation of space group $P\bar{1}$. (From *International Tables*, Vol. I, Kynoch Press, p. 75.)

The logical reaction is to re-choose the axes (Fig. 6*a*) of the unit cell. But this would be inconvenient, since the unit cell could not be described as having $\alpha = \gamma = \pi/2$ (Section 2.4) although it is monoclinic. We therefore retain the monoclinic axes, but state that the crystal has a non-primitive lattice. The symbol for the primitive lattice—the ordinary one—is P; for non-primitive lattices the symbols are A, B or C, according to which face has the centering point, F if all the faces are centred, or I if the centering is at the centroid of the unit cell.

The complete symbol of a space group therefore includes a lattice symbol followed by the symbols for the symmetry elements. An important space group, for example, is $F4/m3m$, since it is the space group of the first crystal structure, $NaCl$[2], and is also that of many of the metallic elements. An important space group for organic compounds is $Pnma$.

2.8.   Representations of space groups

We can represent space groups in two ways: we can give the symmetry elements in their relative positions in the unit cell, or we can give the coordinates of points that are related by these symmetry elements. Thus the space group $P\bar{1}$ can be represented by Fig. 6(b), in which the small circles represent the centres of symmetry (or centres of inversion), or by the *equivalent points* (x, y, z) and $(-x, -y, -z)$. Other space groups are more complicated, and we shall not discuss the matter further, except to say that in the highest symmetry 192 equivalent points are involved.

3.   X-RAY DIFFRACTION

The complete theory of symmetry and space groups is now tied up closely with X-ray diffraction, although it was developed before the nature of X-rays was known. Why do X-rays loom so large in the subject? To understand the answer to this question, we have to recall that, in 1912, the nature of X-rays was still a subject of great controversy in physics: were X-rays waves or particles? (We now know that there is no clear distinction.) The crucial experiment[1] that provided an answer was based upon the guess that a crystal might act as a three-dimensional diffraction grating; a fine beam of X-rays was allowed to fall on a crystal and a pattern of spots was formed on a photographic plate.

This was one of the most important experiments in physics, for it had three major consequences:

(i) It showed that X-rays were waves.
(ii) It provided Moseley[4] with a means of establishing quantitatively the concept of atomic number.
(iii) It provided physics with a radiation short enough to examine matter on the atomic scale. This is the aspect with which this chapter is particularly concerned.

3.1.   The crystal as a diffraction grating

The next crucial experiment after that of Laue *et al.* was the discovery of line spectra by W. H. and W. L. Bragg,[5] who devised and used the ionisation spectrometer. With the powerful tool of monochromatic X-rays, crystals could be examined much more clearly than with the mixture that comes directly from an X-ray tube.

The idea of a single wavelength of X-rays enabled W. L. Bragg[6] to put forward a simplified view of the problem of diffraction of X-rays by a three-dimensional grating. Laue had tried the direct approach of extending the ordinary one-dimensional theory to three dimensions; this, however, was too far ahead of its time and proved sterile as an aid to interpreting

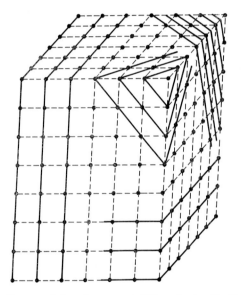

Fig. 7.   Lattice of a crystal showing several different sets of lattice planes. (From Bunn, *Chemical Crystallography*, Clarendon Press, 1961, p. 19.)

X-ray diffraction photographs. Bragg used the simpler approach of regarding a crystal as being composed of inter-penetrating, one-dimensional gratings in many different orientations. To follow this idea more closely we must make use of the lattice concept again.

Since a lattice consists of equally-spaced points in three dimensions it is possible to draw sets of parallel planes through them in many different ways (Fig. 7). The most clearly defined planes contain many points and are relatively widely spaced; they are parallel to possible crystal faces (Section 2.2). The less clearly defined planes contain fewer points and are more closely spaced. We can specify a set of planes by the indices (*hkl*) of the corresponding crystal face.

Bragg's idea was to consider a set of lattice planes as a diffraction grating. Since the absolute position of the lattice is not fixed, we can take a lattice plane to pass through an atom, and then the plane must pass through corresponding atoms in other unit cells. Bragg realised that such a plane should act as a mirror for X-rays in the sense that all the atoms in it must scatter in phase if the angle of incidence equals the angle of scattering (Fig. 8). This result would apply even if the atoms were not regularly spaced in the plane.

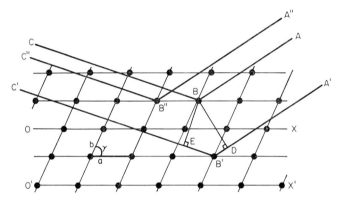

Fig. 8. Representation of radiation scattered from lattice points $BB'B''$. The scattered ray BA will be in phase with the scattered ray $B''A''$ if they both satisfy the reflection condition from the line $B''B$, The scattered ray $B'A'$ will also be in phase if Bragg's law is obeyed.

For appreciable radiation to be scattered, all the planes in a set must scatter in phase, and the condition for this (Fig. 8) is then

$$n\lambda = 2d(hkl) \sin \theta \qquad (1)$$

where $d(hkl)$ is the spacing of the planes. This is Bragg's law, which is of vital importance in the subject; the possible values of θ associated with different hkl's and different n's are known as Bragg angles.

For the more rigorously-minded it is possible to re-state these results in terms of ordinary diffraction-grating theory. For a one-dimensional grating, each order of diffraction is specified by a single integer; for a three-dimensional grating each order is specified by *three* integers, hkl, and these are nh, nk, nl of Bragg's equation.

The intensities of the various orders of diffraction depend upon the arrangement of atoms in the unit cell. If there are N such atoms, the crystal can be considered as composed of N inter-penetrating lattices, each scattering X-rays under the same Bragg conditions, but with different phases. These different phases produce stronger or weaker orders of diffraction, but the complete theory is too extensive to describe here.

3.2. Reciprocal lattice

It is possible to formalise the representation of the diffraction pattern of a perfect crystal by means of a concept known as the reciprocal lattice. By means of this concept, a more complete understanding of the crystallography of materials can be gained, and for such subjects as the theory of metals (Section 5.4) and of imperfect structures (Section 6.3) familiarity with it is essential.

Fundamentally, the reciprocal lattice is the diffraction pattern of the crystal lattice;[7] it is reciprocal to the lattice in the sense that $\sin \theta$ is reciprocally related to d for fixed λ (eqn. 1). The reciprocal lattice consists of a set of points of which the coordinates represent the three integers h, k, l; each point thus represents an order of diffraction.

The complete diffraction pattern of a perfect crystal is represented by the reciprocal lattice, with each point weighted by a quantity proportional to the intensity of the corresponding order.

3.3. Nomenclature

The symbols hkl are used in several different ways in crystallography, and it is therefore necessary to have some conventions concerning their use.

(a) Orders of diffraction are specified by unbracketed symbols, hkl.

(b) Lattice planes and crystal faces are specified by (hkl).

(c) Symmetry-related planes or faces—e.g. all planes such as (110), ($1\bar{1}0$), ($\bar{1}01$), etc., in cubic crystals—are specified by $\{hkl\}$. This symbolism can also be used, where there is no ambiguity, for symmetry-related orders of diffraction.

(d) Directions in a crystal, such as the normal to a face, are specified by [hkl]. A zone axis—a line parallel to a number of crystal faces—is similarly denoted; for example, the zone axis for the faces ($hk0$) is [001].

(e) A group of symmetry-related directions is specified by $\langle hkl \rangle$.

These conventions cover most requirements; any others should be clearly defined when they are used.

3.4. Determination of space groups

We have seen in Section 2.5 that it is not usually possible to determine the true symmetry of a crystal structure from its external form because one cannot distinguish between mirror planes and glide planes, or rotation axes and screw axes. X-ray measurements can enable us to do so, as can be seen from the diffraction-grating theory. Figure 9 shows a one-dimen-

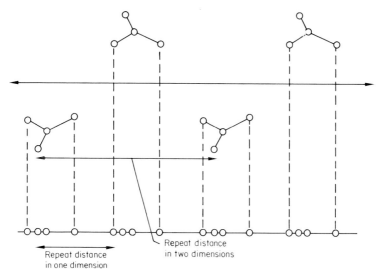

Repeat distance
in two dimensions

Repeat distance
in one dimension

Fig. 9. Structural elements related by a screw axis, showing the projections on a line parallel to the axis.

sional succession of structural elements related by a screw axis. The repeat distance is b, but the projections on to the axis repeat at distances of $\frac{1}{2}b$. Thus, for orders of diffraction from planes perpendicular to the screw axis, the spacing is apparently halved and thus odd orders of diffraction will be absent. Such absences are called systematic absences or sometimes extinctions (although it is difficult to conceive of the extinction of something that has never existed!).

Systematic absences of the type $0k0$ with k odd—those that have just been described—indicate a screw axis. Systematic absences of orders of diffraction with one index zero indicate glide planes and those with no zero index (e.g. hkl when $h + k$ is odd) indicate a centred lattice (Section

2.7). A particularly important example is the systematic absence of orders of diffraction for which h, k and l are of mixed parity—that is, they are not all odd or all even; this indicates the face-centred lattice F.

From the combined observation of external symmetry and systematic absences, one can usually determine uniquely the space group of a crystal. There are some exceptions that need not concern us here. Sometimes, indeed, the systematic absences are sufficient in themselves; for example, if we find no $0k0$ reflections with k odd and no $h0l$ reflections with h odd, we know that the space group is $P2_1/a$ and that the point group (Section 2.3) is $2/m$. Thus all examinations of crystal structures start with a search for systematic absences.

3.5. Structure determination

Having determined the unit cell, the contents of the unit cell and the space group, the problem is finding the positions of the atoms in the unit cell from a knowledge of the X-ray diffraction pattern; this pattern is in the form of a set of numbers $|F(h, k, l)|$ representing the amplitudes of the waves scattered in the various orders (Section 3.1).

There is no straightforward approach to the conversion of a diffraction pattern into a set of atomic coordinates. The problem is exactly the same as that involved in the formation of an image by means of a microscope; but a microscope has lenses that enable the image to be produced by means of the original scattered radiation, whereas X-rays cannot be refracted in the same way. Therefore, the information contained in the *phases* of the scattered waves is lost; in other works, the quantity $|F(h, k, l)|$ gives only the amplitude of each order of diffraction and, to produce an image, the phases have to be inferred in some indirect way. This is the so-called 'phase problem'.

It would be beyond the scope of this chapter to give a detailed account[8] of the various methods of approach to the solution of this problem, but users of results of crystal-structure determination should have some idea of how they are obtained even if only to assess their reliability. The following is a very brief summary.

The oldest method is to postulate a structure and to see if it gives the observed diffraction pattern. It makes use of the structure-factor equation

$$F = \sum_n f_n \exp 2\pi i(hx_n + ky_n + lz_n) \qquad (2)$$

where f_n is the scattering factor of the nth atom in the unit cell and (x_n, y_n, z_n) is its position in coordinates expressed as fractions of the unit cell edges. From the observed intensities the modulus of F the *structure amplitude*, can be derived, and the values calculated from eqn. (2) can be compared with them. Practically all the early work—up to about 1935—was carried out by this method, called the trial-and-error method, but it is limited by the complexity of the structure. When many possibilities of atomic arrangement exist, it is impossible to try them all.

All present-day methods depend upon the process of *Fourier transformation*.[7] The modern formulation of Abbe's theory of image formation is that the image of an object is the Fourier transform of its Fourier transform;[9] the transform of a crystal is its diffraction pattern. For the special case of a periodic structure, the Fourier transform degenerates to a *Fourier series*. To re-transform, the phases must be known and so methods must be devised to find these. For example, the *heavy-atom* method can be used—the introduction into the crystal of an atom heavy enough to be easily found and so to determine the phases of the structure factor. Or we can compare the diffraction pattern of two crystals that are alike except that there are chemically similar atoms of different weight in the two—the *isomorphous-replacement* method. If these methods are not practicable, it may be possible to compare the diffraction patterns of the same crystal taken with radiations of different wavelength—the *anomalous-scattering* method; if the radiation used has a wavelength near to an absorption edge of one particular atom, the scattering factor of that atom may be considerably affected. All these methods have had considerable use in the recent work on proteins, which are, of course, of tremendously greater complexity than would have been considered practicable only a couple of decades ago.

There are also other methods of approach. There is the Patterson method, which involves computing the transform of the intensity—$|F|^2$—distribution instead of the structure-factor distribution, and which gives distances between atoms and not the positions themselves; there are direct methods, which attempt to find phases consistent with a reasonable physical picture of discrete atoms with reasonable separations; and there are optical methods, which attempt to simulate X-ray diffraction patterns in visible light, and so provide an approach more suited to the experimental physicist. A good structural crystallographer will be able to assess the relative possibilities of each of these methods and to use those that he considers appropriate to any particular problems. Despite the lack of a completely direct approach,[8] there is a good probability that any problem can be successfully solved if enough effort is put into it, since computing facilities, which once limited possibilities, are now far more extensive than they used to be.

Notwithstanding all these elaborate devices, it must still be remembered that the determination of a structure is still a physical process, limited by physical considerations. Resolution is limited by the wavelength λ of the radiation used, and cannot exceed $0.3\ \lambda$. The accuracy of the determination of atomic positions is of the order of $\lambda/100$, which by optical standards is very good; claims are often made that $\lambda/1000$ can be achieved, but would need adequate confirmation if they are to be convincing.

3.6. Electron and neutron diffraction

X-rays are not the only radiation that can be used to examine matter on an atomic scale; the theory of de Broglie (1924),[10] that particles with momentum p should behave as waves with wavelength h/p, where h is

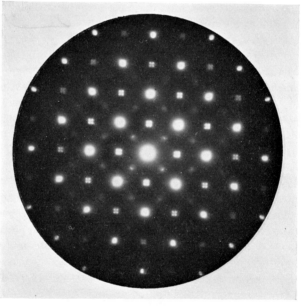

Fig. 10.   Electron-diffraction pattern from a thin film of the alloy $AuCu_3$.

Planck's constant, introduced a variety of new possibilities. The verification of de Broglie's theory by Davisson & Germer (1927)[11] and by Thomson (1928)[12] using electrons accelerated by high potentials showed that these particles could be used and since their wavelengths, given by the equation

$$\lambda = V^{-\frac{1}{2}}1.2 \times 10^{-8}\ \text{cm},$$

could be of the order of 0·1 Å, they should in principle be used to reveal detail below atomic dimensions.

Electrons also have the advantage that their interaction with matter is so strong that electron-diffraction patterns are directly visible on a fluorescent screen and they can be photographed in a few seconds (Fig. 10). But electrons are easily absorbed and so the complete paths must be *in vacuo*. The specimens are limited to thin films or to reflecting surfaces.

More recently the advantages of low-energy electrons have been strongly advocated. Even with potential differences of the order of 100 volts wavelengths of the same order of magnitude as those of X-rays are obtained; such electrons produce what are known as back-reflection patterns. Since electrons are easily absorbed the method is particularly useful for studying surface effects.

Another possibility also exists; electron paths can be modified by electrostatic or magnetic fields in a way that simulates refraction; image formation is therefore possible. Since electrons have short wavelengths, the exciting possibility of images with atomic resolution arises; instruments—*electron microscopes*[13]—have been built to this end. But the aberrations of electron 'lenses' are such that theoretical resolutions cannot be reached and the best resolution claimed is of the order of 3 Å. Individual atoms cannot be seen but large molecules, in regular crystalline array, can be made visible (Fig. 11).

The discovery of neutrons by Chadwick[14] in 1926 introduced another possibility—neutron diffraction.[15] However, strong neutron sources are rare and expensive, and so neutron diffraction can be carried out only in particular places. Nevertheless, much new information has been produced. For example, the scattering factors of atoms for neutrons show no simple progression with atomic number, and so some atoms—such as hydrogen—which are difficult to detect by X-rays are easily found by neutron diffraction. As so often happens in physics, a new discovery opens up unexpected fields; neutron diffraction has made profound contributions to the study of magnetism.

The reason for this is rather remarkable. Neutrons are, in general, scattered only by the nuclei of atoms, which of course play no part in magnetic properties. But neutrons have magnetic moments and these interact with those electrons that are responsible for magnetism, and thus neutrons, as it were, search out these particular electrons and give considerable information about them.

Polarised neutrons—that is neutrons that have a common direction for their magnetic moments, selected by reflection from an appropriate crystal—provide a powerful tool for investigating magnetism; they have uncovered types of magnetism that had previously only been suspected. For example, Néel[16] in 1931, had put forward the idea, from the study of thermal and magnetic properties of materials, that a new type of mag-

Fig. 11.   Photograph of a crystal of tobacco necrosis virus, showing individual molecules. (By courtesy of R. W. G. Wyckoff.)

netism must exist in addition to diamagnetism, paramagnetism and ferromagnetism. In this, the magnetic moments of the atoms alternate in direction and so cancel out, as they do in a paramagnetic material; yet they are perfectly ordered as in a ferromagnetic material. He called this type of structure *antiferromagnetic,* and his ideas have been completely verified by neutron diffraction. In addition, more complicated forms of magnetic ordering have now been found.

These examples show that, with electrons and neutrons, much information complementary to that obtained with X-rays can be produced. The full study of matter requires all three radiations—X-rays for the main work, electrons for surface effects and thin films, neutrons for differentiating atoms with close atomic numbers, for detecting light atoms, and for magnetic studies. Neutron-diffraction work, however, is bound to be confined to a few localities so long as adequate neutron beams can be produced only from the most powerful nuclear reactors.

4. INTERATOMIC FORCES

4.1. Impact of structure determination

From its beginning in 1912, the new subject of crystal-structure determination has affected all branches of science. At last, it was possible to examine matter on an atomic scale, because the radiations used had wavelengths of the same order of magnitude as interatomic distances. It was possible to see how atoms aggregated into solids and so to deduce how they interacted with each other. Some of the earlier speculations were supported, others disproved.

For example, molecules did not exist in inorganic compounds; sodium chloride[2] did not consist of a regular arrangement of NaCl molecules, as the chemistry textbooks said. On the other hand, organic molecules were found to be real entities, and so were radicals such as SO_4 and CO_3.[17] Coordination complexes in hydrates were found, such as $Be(H_2O)_4$ in $BeSO_4 . 4H_2O$ and $Ni(H_2O)_6$ in $NiSO_4 . 6H_2O$, and the functions of water of crystallisation in compounds such as $NiSO_4 . 7H_2O$, $CuSO_4 . 5H_2O$ and $3CdSO_4 . 8H_2O$[18] were explored.

In silicates the importance of the SiO_4 tetrahedron[19] was discovered. Different ratios of silicon to oxygen could be obtained by the sharing of corners and edges of these tetrahedra, and from these considerations it was found that structures were related by geometry, not by chemistry. An explanation of the vast variety of minerals was produced and it provided a classification according to the way the SiO_4 tetrahedra were arranged—in isolated units, in chains in layers, or in three-dimensional networks.

Metals, on the whole, turned out to be surprisingly simple; their structures seemed to be based mainly on packing. But even now we do not know why different elements choose one of the close-packed arrangements—cubic or hexagonal—or the nearly close-packed arrangement known as body-centred cubic. With alloys, however, a more variegated picture emerged; some compositions produced surprisingly complicated diffraction patterns, indicating rather unusual arrangements of atoms.

Organic compounds posed some additional questions. Molecules certainly existed, but how did they hold together when they formed solid compounds? And when low-temperature methods came into existence, and enabled crystals of the inert elements to be produced, what forces held these monatomic molecules together?

All these problems have been tackled by a detailed examination of interatomic distances, and the results have enabled a reasonable self-consistent scheme of interatomic forces[20] to be built up. We do not fully understand all of it, and what we do understand is sometimes too difficult to apply to anything other than very simple structures. But all science starts in this way, and has to be improved upon gradually in the course of time.

4.2. Nature of interatomic forces

The properties of atoms, and particularly the forces of attraction between them, are ultimately based upon their fundamental structures. Since an atom consists of charged particles—the nucleus and an electron cloud surrounding it—electrical forces are possible. The atom as a whole is neutral, however, and other considerations have to be introduced to explain chemical bonding.

The main consideration is the tendency to avoid partially filled energy levels; an atom with an electron short will prefer to acquire an electron from its environment, if this is possible, and to become a negatively charged ion rather than to remain neutral. An atom with a loosely-bound electron will prefer to become a positively charged ion if it can find somewhere to lose an electron to. If two such atoms meet, they can satisfy both their desires; there is a strong electrostatic attraction between the two ions, and this is the nature of the bond—called an *electrovalent* bond—between them.

Another way of avoiding incompletely-filled energy levels is for atoms to unite to share what electrons they possess. Thus, two hydrogen atoms together form a more stable system than each separately, the two electrons providing the stabilising factor. This principle accounts for bonding between similar atoms, and is known as the *covalent* or *homopolar* bond. When several atoms unite to form a molecule the electrons concerned are shared by all the atoms and so are free to move within the molecule.

We shall discuss these ideas, and others that arise from them, in the following sections.

4.3. Ionic bonds

It is clear that ionic bonds will tend to be associated with atoms at the sides of the periodic table—not of course counting the inert gases—since these have most definite electron deficiencies or electron excesses. Thus the alkali metals, with one loosely-bound electron, and the halogens, with energy levels deficient in one electron, form the strongest ionic bonds. Compounds such as NaCl are characterised by hardness and high melting point.

Since the interatomic force is purely electrostatic in nature, it is reduced considerably in a medium with a high delectric constant, such as water.

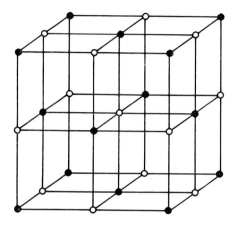

Fig. 12.   Structure of NaCl.

Hence, ionic compounds are soluble in water, in which they dissociate into ions.

The electrostatic force is not directed, and so compounds in which this is the predominant force tend to have structures in which ions of opposite charge take up adjacent positions, each ion attracting as many of the other sort as it can. In NaCl,[2] for example, each sodium atom is surrounded by six chlorine atoms, and each chlorine atom by six sodium atoms (Fig. 12). The actual number of atoms seems to be decided by what we can call the size of the atom—a rather nebulous quantity theoretically, but practically of great importance. For example, the difference between the structures of NaCl and CsCl[18] (Fig. 12) can be regarded as entirely dependent upon the increased size of the caesium atom, which can accommodate eight chlorine atoms around it instead of six.

4.4. Covalent bonds

Covalent bonds are associated with atoms near the middle of the periodic table—C, Si, Ge. The forces are concerned with specific electrons in the atoms, and are therefore directed. For example, the forces that the carbon atom exerts are distributed tetrahedrally, leading in the simplest example, to the well known structure of diamond[18] (Fig. 13). This special distribu-

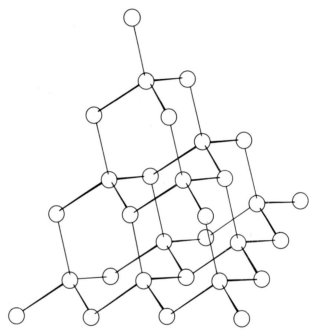

Fig. 13.   Structure of diamond.

tion of carbon bonds leads to the oddly-shaped molecules that organic compounds form and, together with the fact that two or three bonds can combine to form double or triple bonds respectively, leads to the enormous variety of organic molecules that exist.

A chemical molecule can be thought of as an 'atom' with a complicated nucleus and a correspondingly complicated electronic system. For the inner shells the electrons behave more or less as if they were associated with independent atoms, but those in the outer shells take up paths—known as orbitals[20]—related to the complete framework. The theoretical problems involved in calculating these orbitals for any but the simplest and highly symmetrical molecules are extremely difficult.

4.5. The metallic state

When we reach the limit of very large combinations of atoms that are basically simple enough for statistics to be applied, then the picture becomes clearer again, in the same way that the kinetic theory of gases is tractable, although the three-body problem cannot be solved. These conditions apply to metals. We can regard a metal single crystal as a collection of atoms held together by covalent bonds, in which the electrons are shared by the whole crystal. Although we still cannot derive the orbital of any particular electron, we can deal with the average of the whole system.

We can therefore consider a metal as a collection of nuclei and lower-energy electrons in regular array, within a sea of higher-energy electrons. These electrons are called free electrons, and they are responsible for the characteristic properties of the metallic state: because they reflect electromagnetic waves, a metal is opaque and has a typical lustre; because they can move under the action of an electric field, a metal can conduct electricity.

Covalent forces are not directed; thus atoms in metal crystals arrange themselves in forms that involve lowest packing energies. There are two main close-packed planes of atoms; in hexagonal close-packing the planes repeat alternately and in cubic close-packing they repeat at intervals of three layers. Many elements, however, adopt the body-centred cubic structure which is not quite close-packed.

These principles do not, however, apply generally to alloys. Packing becomes of lesser importance and some relatively complicated structures arise. The stability of these structures now depends upon the free electrons, and a lower energy may result from a more complicated structure as we shall show in Sections 5.3 and 5.5. Even some of the elements, such as manganese and tin, have complicated structures. For these elements the electron configurations are rather uncertain, and it may be that these elements should be considered as alloys of atoms with different electron configurations.

4.6. Van der Waals forces

Interatomic forces other than ionic and covalent must exist since organic compounds, for which the covalent forces are fully satisfied within each molecule and which have no ionic forces, can still crystallise. A residual force of some sort must exist, and this is called the van der Waals force. It has no explanation as completely satisfying as electron transfer or electron sharing, and it is now believed to result from transitory dipole moments.

In any piece of neutral matter the centroids of the positive and negative charges should coincide. In fact, because of random variations, it is probable that the centroids will vary, so producing a variable dipole moment. Such a dipole moment will induce dipole moments in neighbouring molecules, and the two will thus attract, in the same way that a magnetic dipole attracts a piece of soft iron. The dipole moments of neighbouring molecules will be continually varying, but the net result must always be an attraction, albeit slight.

The fact that van der Waals forces are weak is responsible for the softness of organic crystals and for their low melting points. The larger the molecules, the larger the possible dipole moment, and so the more likely the compound is to crystallise. For the simple atom helium the force is very weak, and, as is well known, this element is the most difficult to crystallise, being liquid at ordinary pressures down to the lowest temperatures available.

Van der Waals forces between molecules are recognised by interatomic distances of the order of 3·3 Å.

4.7. Hydrogen bonds

In many organic structures, some intermolecular distances appreciably less than 3·3 Å are observed; they may be as small as 2·4 Å but are generally grouped around 2·6 Å. Since they are always associated with the presence of hydrogen atoms, they are called hydrogen bonds.[21] These bonds are still not completely understood, but theory has hardened towards the following explanation.

We have to ask 'What is particular about the hydrogen atom that it should have this special property?' The answer is that it is the only atom that has a completely exposed nucleus when it is ionised. Thus, if its electron has taken part in bond formation the proton can attract certain other elements—oxygen particularly, and also nitrogen and chlorine to a lesser extent.

The hydrogen bond is much weaker than the covalent bond but much stronger than the van der Waals bond. Consequently it cannot disturb molecular shapes appreciably, but it can hold molecules together fairly firmly; consequently crystals with many hydrogen bonds are harder and of higher melting points than crystals held together with only van der Waals bonds. Its property of holding molecules firmly, yet gently, is of paramount importance and may be, as we shall see in Section 6.4, one of the chief factors in making living matter possible.

4.8. Bonds in crystalline matter

Although we have made a definite classification of bonds into the various types possible, in practice they rarely fall precisely into the categories described. In the alkali halides it is probable that the bonds are almost entirely ionic, in organic molecules the bonds are almost entirely covalent, and in hydrocarbons the intermolecular forces are almost entirely van der Waals. Generally, it is possible that a given bond may be partly ionic and partly covalent; a parameter of this sort is necessary in order to account for certain systematic variations in interatomic distances in related compounds.

Few compounds have bonds only of one type. For example, a hydrated inorganic salt will have covalent bonds between the metal atom and the water molecules, ionic bonds between the metal–water ion and the acid radical, and possible hydrogen bonds between the water molecules. For example, $NiSO_4.7H_2O$ has a complex of $Ni(H_2O)_6{}^{2+}$ formed by covalent forces and a complex, $SO_4{}^{2-}$, also formed by covalent forces; the two ions have ionic forces between them; the extra water molecule is hydrogen-bonded to the other water molecules and seems to act merely as an extra link in the structure to enable a more stable configuration to grow.

In silicates, as we have seen, the most important element is the SiO_4 tetrahedron,[19] which is held together by covalent bonds. This unit can join to other tetrahedra by sharing one oxygen atom or by sharing two; this is a type of polymerisation. This process can continue by adding more tetrahedra in one, two or three dimensions. The negative ions so formed are joined by ionic forces to the metal atoms that are needed to produce a neutral structure.

The most electronegative ion is SiO_4 itself, since it has the largest ratio of oxygen to silicon. Thus, as corners and edges are shared, fewer metal atoms of any particular type are needed for stability; in this way a very large variety of silicate minerals can occur. The metal atoms fit into interstices between the oxygen atoms and the way in which they fit in seems to be dictated almost completely by atomic size, again showing the importance of this quantity; smaller atoms, such as beryllium and to some extent aluminium, fit into tetrahedral interstices, whereas larger atoms such as sodium and potassium fit into octahedral interstices.

In SiO_2 itself the principle is taken to extreme; the whole arrangement of tetrahedra is three-dimensional and is electrically neutral so that no further ions are necessary for stability. Several different forms of SiO_2 are known, their formation being dependent upon temperature and pressure; they are all very hard and of high melting point as all covalent compounds are.

4.9. Summary

Although the description of chemical bonds in the terms outlined in the last few sections is extremely useful, it must not be accepted as complete. Each type of bond is merely a well-defined stage in a larger continuous scheme which must be based on the wave function of the complete arrangement of atoms; a particular conformation exists because its wave function leads to a lower free energy than any other. But we are very far from being able to derive a general treatment with enough rigour to solve any problem except those of exceedingly simple molecules.

As molecules become larger, so presumably the nature of the forces becomes more finely graduated and so more difficult to place in the conventional classifications. For example, what type of force holds the oxygen molecule to the haemoglobin molecule? The force is weaker than the hydrogen bond and stronger than the van der Waals force; it is pressure-sensitive and so provides the breathing mechanism that makes life possible. In our present state of knowledge it appears that the derivation of the wave function of the haemoglobin molecule presents insuperable problems. It is probable that it will never be solved at all.

For this sort of reason, it is likely that the present system of interatomic forces—crude as it is—will have to satisfy us for a long time, as a basis for understanding chemical and other processes.

5.   METALS AND SEMICONDUCTORS

5.1. Characteristics of metals

The greatest success in the application of electronic-bond theory lies in the field of metals; metal atoms are simpler than chemical molecules, and a metal crystal contains so many atoms that statistical methods can be applied to them. The discrete energy levels in a single atom, which become more complicated in a chemical molecule, merge into a continuous function as far as the outer electrons in a set of metal atoms are concerned. And it is these electrons that characterise the metallic state.[22]

In everyday experience it is quite easy to make a clear distinction between metals and non-metals, although in detail many difficulties arise. Metals have the characteristics of being strong, and good electrical and thermal conductors; in particular the difference between the electrical conductivity of metals and non-metals, which is many orders of magnitude, makes the distinction easy, and it is only recently that the study of materials such as semiconductors has arisen to blur the dividing line. But as an example of how precarious this line is, we should remember that the element tin has two allotropic forms—white tin which is a metal, and grey tin which is a non-metal.

The position of metallic elements in the periodic table is well defined; they are all on the left-hand side, indicating that metallic atoms have an outer shell containing only a few electrons. The best conductors are those metals, such as copper and sodium, which contain one electron per atom, or aluminium, which contains three. Elements such as beryllium, graphite, sulphur and diamond, which contain two or four electrons per atom, are either poorer conductors or even insulators. Such facts must form a basis for the theory of metallic structure.

A further important observation about metals is that the best conductors are pure metals. Although alloys may still have well-defined crystal structures, they are almost invariably worse conductors than their pure constituents.

5.2. Electronic structure of metals

The basic characteristic of metals is that they contain free, or almost free, electrons. The few electrons in the outermost shell in the isolated atom are no longer bound to their parent atom but are free to wander throughout the metal, subject only to the restriction that they do not accumulate in a localised region to produce a space charge. It is this freedom of motion which gives rise to the high electrical conductivity and also to the high cohesion in metals. But why is there such a big difference between metals and non-metals in these respects? The answer is quite simple.

In an isolated atom, the electron at distance r from the nucleus has a potential energy e^2/r (Fig. 14(a)). When several atoms come together to form a regular crystal, the potential energy seen by an electron becomes the superposition of such a potential centred on each atom (Fig. 14(b)). In this periodic potential, the potential wells around the atoms are not quite as deep as in the lone atom. Thus an electron of energy E_1, which is tightly bound, is hardly affected by the neighbouring atoms; such electrons are called *core electrons* and are relatively unimportant in the theory of electronic structure. A loosely-bound electron, of energy E_2, is affected greatly by the fact that its atom is now a member of the crystal. In the lone atom it was bound, but in the crystal it is free to wander throughout the metal with very little hindrance from the atoms. Such *almost-free electrons* are important in conductivity and cohesion. If the energy is about E_3, where it has to tunnel quantum-mechanically through every barrier, the electron is able to move, but only sluggishly, through the metal.

In an irregular array of atoms, or in a disordered alloy, the range of energies requiring tunnelling is greatly extended (Fig. 14(c)). and hence the lower conductivity follows. The regular crystal structure is of vital importance to high conductivity.

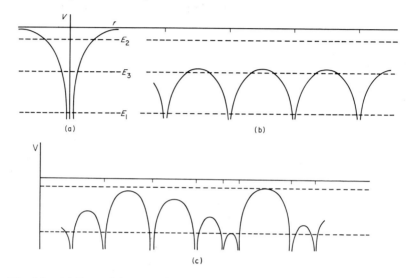

Fig. 14.   (a) Potential energy in an isolated atom; (b) Potential energy in a periodic crystal; (c) Potential energy in a disordered crystal.

5.3.   Cohesion of metals

The cohesion of a metal depends on two factors—the cohesion of the positive metallic ions, and the contribution from the free, or nearly-free, electrons. It is easy to see that almost all the cohesion arises from the latter. As, on average, the positive lattice charge and the negative electronic-charge distributions cancel one another out, it is approximately correct to consider the positive ions as having a cohesion equal to the equivalent inert gas. For example, sodium has an ion with the electronic structure of neon. As we know, the cohesive energy of neon is so low that it melts at $24°K$ and vaporises at $27°K$. Clearly, the free electrons in sodium are mainly responsible for the cohesion, leading to a melting point of $372°K$ and boiling point of $1156°K$. The comparison is not so easy to make for copper, because the inert-gas equivalent to its ion does not exist, although its properties can be imagined without too much extrapolation.

A further point about electronic cohesion is clear from any table of melting-points and boiling-points of the elements. The melting-points and boiling-points of non-metals are generally fairly close, indicating that the cohesive energy of such elements is strongly dependent on the existence of a crystal structure. For metals, however, there is generally a large difference between melting-points and boiling-points, showing that electronic cohesion is not so dependent on the crystal structure, and does not break down in the amorphous liquid state. Nevertheless, the electronic cohesion is affected to a certain extent by the crystal structure, otherwise sodium would have the same melting point as neon even if its boiling point were still 1156°K.

It is impossible to outline the theory of electronic structures of metals in a short article. We can, however, briefly mention the basic factors

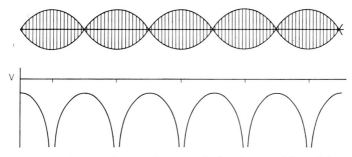

Fig. 15.   Potential energy in a perfect crystal, showing matching of the energy with an electron wave.

influencing cohesion and crystal structure. Since electrons must be treated as waves on a microscopical scale, it is easy to see that the total energy of the structure can be lowered if as many electrons as possible have wave-functions such that their maximum probability densities coincide with the minima of potential arising from the periodic lattice of ions. If the electron waves have wavelength λ, the probability density, being the square of the wave function, also has this periodicity, and the minimum total energy occurs when λ and the lattice spacing are equal (Fig. 15). Of course, not all the electrons can have this favourite wavelength because the Pauli exclusion principle limits the number of electrons in any one state to two (of opposite spins) and λ acts as a label to the state. But by making λ the minimum wavelength available (the maximum energy) the largest possible number of electrons can achieve it and so the metal has minimum energy. Thus we have achieved cohesion in the metal. Now λ depends mainly on the density of the metal, and only marginally on its crystal structure. The structure can now adjust itself to take maximum advantage of λ by providing a periodicity of λ in as many directions as possible. It cannot achieve this in every direction, because of the necessity for translational symmetry; but will do its best, and this may warrant a crystal structure that is quite complicated.

5.4. Fermi surfaces and Brillouin zones

Over the years, a very helpful pictorial representation[22] has evolved, and, in many people's minds, has almost achieved a reality of its own. This representation involves a new type of space called *K space,* which is geometrically similar to reciprocal space (Section 3.2). A complete descrip-

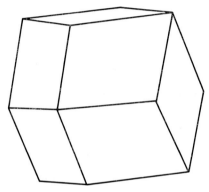

Fig. 16.   Brillouin zone of the body-centred cubic structure.

tion of the electrons in a crystal would require six-dimensional space—three space coordinates and three components of momentum; in *K* space we deal only with the latter, and it is therefore sometimes called *momentum space* or *phase space.*

In *K* space each electron is represented by a point whose coordinates are the three components of momentum, *p.* If it were not for Pauli's exclusion principle, all the electrons would be near the origin of *K* space, since the states there represent the lowest energies; Pauli's principle—which applies to the whole crystal as much as to a single atom—ensures that they are well spread out. It is the electrons at the higher levels in *K* space that give solid matter its most interesting properties.

The surface bounding the occupied states is known as the *Fermi surface.* If the electrons were free, the Fermi surface would be spherical, since energy is $p^2/2m$, where *m* is the electronic mass. However, the interactions mentioned earlier produce some interesting distortions of the Fermi surface, since they lead to the result that the energy is not always $p^2/2m$.

This result arises because electrons with certain momenta—and consequently wavelength—can be reflected from lattice planes in the crystal. It can easily be shown that the points in K space representing 'reflectable' electrons lie on planes. Detailed theory shows that states just within these planes have energies less than $p^2/2m$, but at each plane a discrete energy increase occurs, and electrons outside have energies greater than $p^2/2m$.

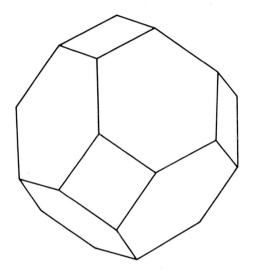

Fig. 17.   Brillouin zone of the face-centred structure.

Thus it pays a crystal to distort its Fermi surface so that it approaches the planes; but it must avoid, if possible, breaking through a plane, as a discrete increment of energy is involved.

These planes outline polyhedra known as Brillouin zones: the first zone for the body-centred cubic structure (*e.g.* α-Fe) is composed of planes corresponding to the 110 reflections (Fig. 16); for the face-centred cubic structure (*e.g.* Cu) it is composed of planes corresponding to the {111} and {200} reflections (Fig. 17). The complicated structure mentioned at the end of Section 5.3 can be regarded as satisfying the condition that the Fermi surface should touch the Brillouin zone at as many points as possible.

5.5. Electron–atom ratios

The property by means of which this end is achieved is the ratio, e/a, of the number of valency electrons per atom. Thus metals and alloys with the same value of e/a tend to have the same structure, a rule that was pointed out by Hume-Rothery[23] and which is known by his name.

For pure metals e/a must normally be integral, but for alloys it can take arbitrary values and so many more structures exist for alloys than for metals. One particular alloy structure of common occurrence is the γ structure of which the prototype is Cu_5Zn_8;[24] its Brillouin zone (Fig. 18)

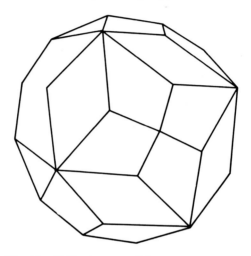

Fig. 18.   Brillouin zone of the gamma structure.

is defined by the reflections {330} and {411} and is nearly spherical. The structure of β-manganese,[25] with 20 atoms in the unit cell, may well be considered as an alloy of two forms of manganese atoms with different electronic structures (*see* Section 4.5); several alloys (*e.g.* Ag_3Al) with e/a equal to 1·5 adopt this structure.

5.6. Conduction

Brillouin-zone theory accounts beautifully for electrical conduction, explaining clearly how solids can be divided into conductors and insulators, with very few materials in between.

The action of an electric field is to tend to make the free electrons move— that is to make the Fermi surface asymmetric (Fig. 19) since the number of occupied states with momentum along the electric field must be reduced, and the number against the field must be increased. (Remember that the electronic charge is negative.) This is easy if the Brillouin zone is only half-full, as it is for elements with one electron per atom, or with three, which overlap into the second zone. If the metal has two electrons per atom, the zone will be nearly full, and so change of state will be more difficult. Thus metals with one or three electrons per atom are better conductors than those with two (Section 5.1).

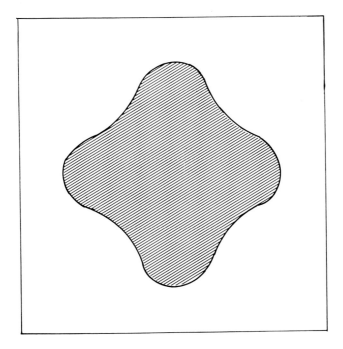

Fig. 19(a)—see over

Suppose, however, that a Brillouin zone is quite full. To produce electron drift, the states would have to break through the zone surface, which would need a disproportionately large increase in energy. Such solids are insulators.

One can simulate an increasing occupancy of a Brillouin zone by adding a trivalent metal, such as aluminium, to a monovalent metal such as copper. When the number of electrons per atom reaches two, should not the material become an insulator? It does not, because it turns out that the states of highest energy within the zone—that is in the sharp corners (*see*

Figs. 16–18)—are higher than the states of lowest energy outside. Thus, for lowest electronic energy, the first Brillouin zone will not fill up before the occupancy of the second starts. Thus there is always room for electron drift and the material remains metallic.

Potential gradient

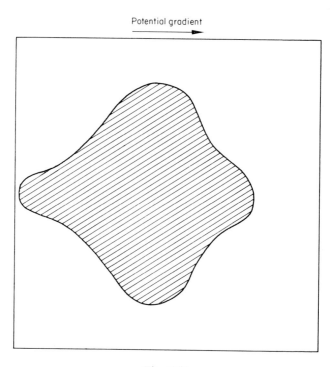

Fig. 19(b)

Fig. 19. Two-dimensional representation of a Fermi surface in a crystal (*a*) without a potential gradient, (*b*) with a potential gradient. (Effect greatly exaggerated.)

5.7. Semiconductors

The distinction between conductors and insulators is thus quite sharp and there would not appear to be any scope for materials with intermediate properties. We have, however, neglected the effects of temperature, which, since it increases the random energy of the electrons, blurs the Fermi surface; but this blurring is extremely small, even up to the temperature at which the most refractory solid melts.

Suppose, however, that the energy gap between the maximum in the lowest Brillouin zone and the minimum in the next is very small—comparable with the thermal energy. Then some electrons might occupy states in the upper zone, leaving spaces in the lower one; weak conduction then becomes possible. This is semiconduction,[26] and takes place in elements such as silicon and germanium, which are both tetravalent.

The characteristic of semiconduction is that it *increases* with temperature. Ordinary conduction decreases with temperature because the irregularities produced by the thermal motions of the atoms as a whole tend to impede the electron flow. In semiconductors, the extra spaces made available in the lower zone when the thermal energy puts electrons into the upper zone, makes conduction easier as temperature increases.

With energy differences in semiconductors so nicely balanced, slight variations can have enormous effects. A few foreign atoms of different valency—phosphorus which is pentavalent or boron which is trivalent—in crystals of silicon or germanium give energies that lie within the forbidden gap and allow fine control of electronic states with very small fields. These are the materials of which transistors are made, but the complete theory is too extensive to be included here.

6. STUDY OF ORDINARY MATTER

6.1. General principles

So far we have concerned ourselves only with perfect crystals. It has turned out that the study of these materials has led to a great increase in our understanding of the forces holding solid matter together and has also introduced new devices of extreme technological importance. We must remember that ordinary matter does not consist of perfect crystals: very little of it has a high degree of purity, although its crystallisation is quite marked; some solid material has practically no crystallinity at all.

For studying these materials, the theory we have so far outlined is not wasted; in fact it forms a necessary basis for adaptation to the problems that arise. For example, the existence of the reciprocal lattice (Section 3.2) implies perfect crystallisation: for imperfect crystals we have to extend the concept, first by relaxing the condition that the reciprocal lattice should consist only of points and then by considering effects between the reciprocal-lattice 'points'—that is, in the whole of what we call *reciprocal space*.[27]

The more imperfect a crystal, the greater the invasion of the diffraction pattern into reciprocal space. If there is no periodicity at all, then the diffraction pattern covers the whole of reciprocal space. It is clear that studies of such diffraction patterns are bound to be more complicated than those of perfect crystals, and we can do no more than outline the general principles here.

6.2. Metals

A pure metal, carefully annealed, can be remarkably perfectly crystalline, consisting of a mass of small crystals, randomly oriented. In practice such a piece of material is not of much use; most of the metallic articles that we use have had their properties improved by the deliberate introduction of imperfections.

The simplest method of introducing imperfections is distortion, which may occur in the process of fabrication—rolling sheets or drawing wires, for example. Two effects can be envisaged; the crystals may be broken up into much smaller ones or they may be twisted into irregular shapes. Both effects would result in broadening of the X-ray reflections and this is observed experimentally. Which effect, then, *does* happen?

We can sort out the two effects by considering the modification to the reciprocal lattice introduced by each in turn. A small crystal would have a reciprocal lattice consisting of broadened 'points'; a distorted crystal could be considered as having planes of variable spacing and thus also produce broadened 'points'. But the former effect produces 'points' all of the same size (Fig. 20(a)), whereas the latter produces 'points' whose extent is proportional to distance from the origin (Fig. 20(b)).

It is not as easy as it appears to differentiate between these two effects, particularly when X-ray patterns can be obtained only of polycrystalline specimens. It turns out that the distortion picture is nearer the truth; the crystal grains are broken up somewhat, but they are also distorted into irregular patterns.

We can build up a picture of the deformation of a metal. When the crystals are perfect, the planes of atoms can glide easily over each other and the metal is mechanically weak. When the crystals are distorted, relative movements of the planes become more difficult, and the metal becomes harder.

Temperature provides an opposing mechanism; thermal vibration will cause the atoms to try to regain their perfect crystalline arrangement and thus the nearer a metal is to its melting point, the softer it is likely to be, since large strains cannot persist in it. Lead is a much softer metal than iron, for example.

It would be helpful if distortions could be introduced into some metals

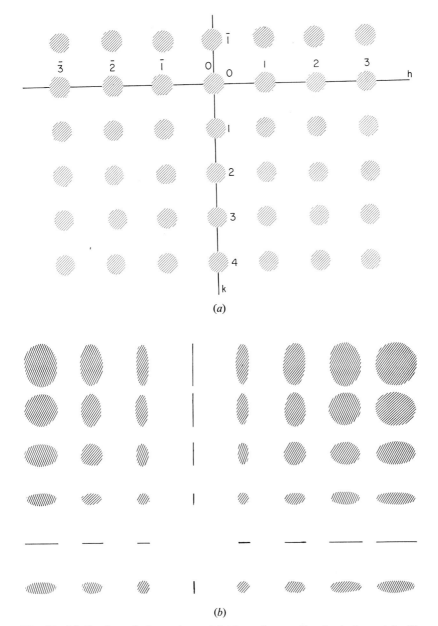

Fig. 20. (a) Section of the reciprocal lattice of a small spherical crystal; (b) Section of the reciprocal lattice of a distorted crystal.

of low melting point to produce a greater hardness. Aluminium, for example, is light and resistant to corrosion; its melting point (660°C) ensures that it is rather soft for most purposes. Defects can, however, be inserted artificially. If 4 per cent of copper is introduced, appropriate heat treatment can cause platelets of composition $CuAl_2$ to be deposited. (Fig. 21); these platelets act as 'keys' to prevent easy glides of the atomic

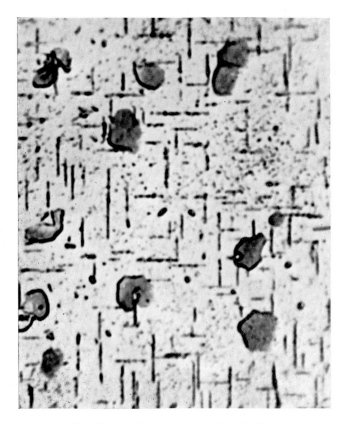

Fig. 21. $CuAl_2$ platelets in Al–Cu alloy.

planes and so the material—duralumin—is much harder than pure aluminium.

The first evidence for the existence of these platelets lay in the X-ray photographs of a single crystal of the alloy.[28,29] These showed streaks (Fig. 22) which indicated that platelets of predominantly larger atoms were present parallel to the {100} planes. During the course of heat treatment these streaks developed into spots and finally could be interpreted as belonging to a definite structure of crystallites. By this time, the hardness produced by the 'keys' had been greatly reduced.

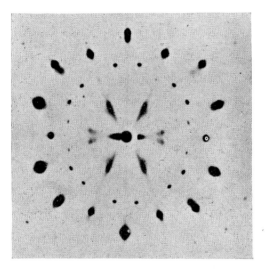

Fig. 22. Single-crystal photograph of duralumin showing streaks. (From *Research*, 1949, **2**, (5) 209, Fig. 9.)

By careful control of time and temperature the maximum hardness can be produced, and this is responsible for the very extensive use of aluminium in our present-day life.

6.3. Polymers

One of the most remarkable chemical achievements of recent years has been the discovery of methods of making large molecules out of small ones. A small molecule like ethylene can be made to combine with other molecules to form a chain of indefinite length—a polymer, polyethylene (or polythene). In this way, a solid is made from a gas.

Since the molecules are of indefinite length, they do not readily crystallise, although they have some crystalline properties. X-ray diffraction photographs show rather fuzzy spots on a general background (Fig. 23), and there have been many discussions about the information that these photographs convey. To the crystallographer, who has absorbed the principles of single-crystal theory, the diffuse spots and the general back-ground indicate the presence of crystalline and amorphous material respectively.

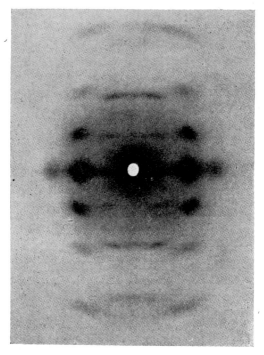

Fig. 23. X-ray diffraction photograph of natural silk. (From *Textile Fibres under the X-rays,* ICI, 1941, p. 22, Fig. 3.)

The theory, however, is not self-consistent. Attempts to measure the ratio of amorphous to crystalline matter by other methods show no agreement at all and it is fairly certain that, when the complete structure of polymeric materials does emerge, it will be something much more compli-cated than a simple mixture.[30]

It is probable that the diffraction pattern must be treated as a single entity. From it we can plot the distribution of scattering in reciprocal space (Section 3.2) and from this we should be able to derive a scheme of atoms that could produce it. The task is much more difficult than that of working out a crystal structure, since we have to account for the whole of the intensity in reciprocal space, not merely the intensities at specific points. But undoubtedly this is one of the tasks to which X-ray diffractionists should attach themselves in the future.

A theoretical difficulty exists, however, that may make the investigation of imperfect structures more difficult than has so far been realised. In Section 3.5 it was pointed out that an image is formed by two processes of Fourier transformation; the first is the scattering function of the object and the second is the collection of the scattered radiation to produce an interference pattern which is the image.[31] A great deal of attention has been given to the second process, which involves the design of complicated optical systems—microscopes and electron microscopes (Section 3.6); but the first process is left to look after itself. But it is no use transforming a function that is not a reasonably accurate representation of the Fourier transform of the object and we must ask ourselves how far the scattering function satisfies this condition.

The trouble arises because radiation can be multiply-scattered. The radiation scattered in any one direction is therefore not a simple function of the structure of the specimen, but is made up of the directly scattered waves combined with other scattered waves that have been re-scattered into the same direction. To simplify the problem, we have worked out the complete scattering function of two exactly similar plane objects separated by a distance l; it then appears that the complete function is an adequate representation of the Fourier transform of the single object only over an angle θ_m of the order of $\sqrt{\lambda/2\pi l}$, where λ is the wavelength of the radiation. The limit of resolution Δ corresponding to this angle is given by the well-known expression

$$\Delta = \lambda/\sin\theta_m$$

$$\simeq 2(\lambda l)^{\frac{1}{2}} \text{ if } \theta_m \text{ is small.}$$

For electrons with $\lambda = 0.05$ Å and a specimen of thickness 1000 Å, this expression leads to a resolution of about 15 Å, which is greater than the limit, 3 Å, that is being currently claimed. It may well be, however, that this fine detail is not a true representation of the structure of the specimen. For resolutions of the order of 3 Å, the specimen would have to be not more than 50 Å thick.

Another deduction from the theory is that the limit of resolution of the electron microscope is proportional to $V^{-\frac{1}{4}}$ where V is the operating voltage. Thus increasing the voltage makes only a marginal difference to the resolution—a fact that has been observed in recent research on the new megavolt electron microscopes.

The type of radiation is not specified in the theory. Why, then, does it not apply to X-ray diffraction? We know that the X-ray diffraction pattern of a crystal *is* an accurate representation of its transform. The answer may lie in the fact that for a perfect crystal the conditions for even double diffraction are very precisely defined,[32] and it does not often occur. Even so, for very precise work it should be taken into account, although it rarely is so.

For non-crystalline matter multiple diffraction will take place more readily, and so it may be that the diffraction patterns of such materials as polymers are not adequate representations of their Fourier transforms. This fact may explain the difficulties that have been experienced in interpreting such diffraction patterns. If so, the theory pin-points the problem: specimens must be as thin as possible. Progress in the subject may well depend upon techniques for making very thin specimens and for obtaining observable diffraction patterns—in X-rays or electrons—from them.

6.4. Living matter

Success in determining atomic arrangements in polymers may lead to a much more distant goal—the study of living matter itself. Some parts of the body, such as the teeth and bones, are—relatively speaking—fairly well crystallised, but others, such as hair and muscle, give very diffuse X-ray diffraction patterns. Nevertheless, it is surprising how much information has been obtained about the way that living matter is built up. It may be that some of the deductions are merely daring speculation, but if such speculation gives future workers a basis to build on, it will have been worth its while.

A more definite approach has been to try to induce perfect crystallisation in the components of living processes. The most outstanding example has been the preparation of crystals of haemoglobin and myoglobin—constituents of the blood. The problems of crystal-structure determination were immense, but the methods outlined in Section 3.5 were ultimately, completely successful.[33] We now know the way that the atoms are arranged in the molecules that make life possible.

This information provides only a beginning. Why do the atoms arrange themselves so? What is the nature of the force holding the oxygen molecule so gently, yet firmly, to the haemoglobin molecule in the breathing process of life? These, and a multitude of other questions, are bound to flow from the increasing penetration devoted to the study of living matter. And what we do uncover seems to be so much more complicated than that which was expected.

7. SUMMARY

We have tried to show in this chapter how the fundamental study of crystallography has its impact on almost every branch of science. Whenever solid matter is involved, its examination on an atomic scale requires a short-wave radiation such as X-rays, and a thorough knowledge of its diffraction by perfect crystals. From this we can try to extend the scope of the method to less perfectly crystalline matter, to the deformed crystals that form the materials of our civilisation, down to the molecules of life itself.

We have come a long way since the first X-ray diffraction photograph was taken in 1912!

8. ACKNOWLEDGEMENT

We are grateful to Mrs Judith M. Cohen (née Lipson) for her help in preparing the diagrams for this chapter.

9. REFERENCES

1. Ewald, P. P. (1962). *Fifty Years of X-ray Diffraction*, International Union of Crystallography, Utrecht.
2. Bragg, W. L. (1913). *Proc. Roy. Soc.*, **A89**, 248.
3. *International Tables for X-ray Crystallography I*, Kynoch Press, Birmingham, 1952.
4. Moseley, H. G. J. (1913). *Phil. Mag.*, **26**, 1024; *Phil. Mag.* (1914), **27**, 703.
5. Bragg, W. H. and Bragg, W. L. (1913). *Proc. Roy. Soc.*, **A88**, 428.
6. Bragg, W. L. (1912). *Proc. Camb. Phil. Soc.*, **17**, 43.
7. Lipson, H. and Taylor, C. A. (1958). *Fourier Transforms and X-ray Diffraction*, Bell, London.
8. Lipson, H. and Cochran, W. (1966). *The Crystalline State*, Vol. III, Bell, London.
9. Taylor, C. A. and Lipson, H. (1964). *Optical Transforms*, Bell, London.
10. Broglie, L. de (1924). *Phil. Mag.*, **47**, 446.
11. Davisson, C. J. and Germer, L. S. (1927). *Nature*, **119**, 558.
12. Thomson, G. P. and Reid, A. (1927). *Nature*, **119**, 890.
13. Cosslett, V. E. (1950). *Introduction to Electron Optics*, Clarendon Press, Oxford.
14. Chadwick, J. (1932). *Nature*, **129**, 312.
15. Bacon, G. E. (1962). *Neutron Diffraction*, Clarendon Press, Oxford.
16. Néel, L. (1932). *Ann. Phys.*, **17**, 64.
17. Bragg, W. H. and Bragg, W. L. (1933). *The Crystalline State*, Vol. I, Bell, London.
18. Wyckoff, R. W. G. (1953). *Crystal Structures*, Interscience, New York.
19. Bragg, W. L. and Claringbull, G. F. (1966). *The Crystalline State*, Vol. IV, Bell, London.
20. Pauling, L. (1960). *The Nature of the Chemical Bond*, Cornell University Press, Ithaca, New York.

21. Bernal, J. D. and Megaw, H. D. (1935). *Proc. Roy. Soc.*, **A151**, 384.
22. Mott, N. F. and Jones, H. (1936). *The Theory of the Properties of Metals and Alloys*, Clarendon Press, Oxford.
23. Hume-Rothery, W. (1926). *J. Inst. Metals*, **35**, 295.
24. Bradley, A. J. and Thewlis, J. (1926). *Proc. Roy. Soc.*, **A112**, 678.
25. Bradley, A. J. and Thewlis, J. (1927). *Proc. Roy. Soc.*, **A115**, 456.
26. Shockley, W. (1950). *Electrons and Holes in Semiconductors*, Van Nostrand, New York.
27. Lipson, H. (1947). *Institute of Metals Monograph*, p. 35.
28. Preston, G. D. (1938). *Nature*, **142**, 570.
29. Guinier, A. (1937). *Compt. rend.*, **204**, 1115.
30. Peiser, H. S., Rooksby, H. P. and Wilson, A. J. C. (1955). *X-ray Diffraction by Polycrystalline Materials*, Institute of Physics, London.
31. Lipson, S. G. and Lipson, H. (1969). *Optical Physics*, Cambridge University Press.
32. Renninger, M. (1937). *Z. Krist.*, **97**, 107.
33. Perutz, M. F. (1962). *X-ray Analysis of Haemoglobin*, Les Prix Nobel, Stockholm.

CHAPTER 2

DEFECT STRUCTURES

A. G. CROCKER

1. INTRODUCTION

The hypothesis that crystalline materials consist of regular arrays of atoms, ions or molecules leads to a satisfactory understanding of some physical and mechanical properties, such as specific heats and elastic constants. It cannot, however, explain many other aspects of the behaviour of materials, particularly the notorious weakness of metal single crystals and the electrical properties of ionic crystals. It is therefore natural to postulate the existence of defects in crystal structures. These may be envisaged to have zero, one, two or three dimensions and can thus be conveniently labelled point-, line-, sheet-, and volume-defects respectively. All four types of defect are in fact known to occur in crystalline materials and to control many of the important properties. They are also intimately related to each other so that, for example, the boundary of a volume defect may be considered to be composed of sheet defects, consisting of networks of line defects, which attract or repel point defects. More directly, volume defects may arise through the clustering of point defects. Relationships and interactions of this kind are discussed at appropriate places in this chapter, but it has been found convenient to make a basic division of the text according to the main type of defect being considered. Thus, Sections 2 to 5 deal primarily with point, line, sheet and volume defects respectively; some concluding remarks are given in Section 6. The emphasis throughout is on the structure of the defects, so that geometrical considerations predominate. Some basic general theory is presented, particularly for the case of line defects, but wherever possible the restrictions imposed by specific crystal structures and characteristic interatomic bonding forces are described. In particular, defects in pure metals, ordered and disordered alloys, ionic and covalent crystals, semiconductor materials

47

and crystalline polymers, covering a variety of different crystal structures, are discussed. No attempt is made to give detailed descriptions of the experimental methods used to observe defects but the micrographs which have been included illustrate most of the important techniques.

2. POINT DEFECTS

The simplest type of crystal defect which can be envisaged is the point defect, typical examples being the vacant lattice point, or vacancy, the

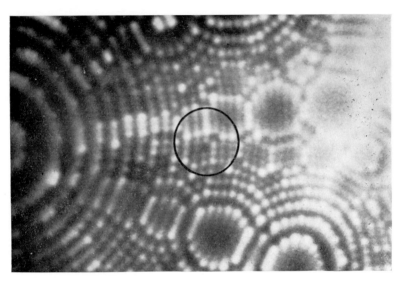

Fig. 1.   Field ion micrograph showing vacancies in an irradiated Ir specimen. The bright spots indicate the positions of atoms at the perimeters of low index planes which intersect the hemispherical tip of the specimen. Sets of parallel planes produce the series of concentric rings. Two atoms are missing at the centre of the encircled region. Courtesy M. M. B. Fortes and B. Ralph.

intrusive atom not situated at a lattice point, or interstitial, and the impurity atom. However, due to the atomic dimensions of these defects, experimental information on their structure and behaviour has been primarily indirect. Thus, for example, measured changes in electrical and thermal resistivity, density and internal friction, due to assumed variations in defect concentrations, have all provided valuable information, but the detailed models of the configurations of many point defects are still speculative. Individual atoms, and hence point defects, can now be resolved however, using the field ion microscope; a micrograph showing vacancies in irradiated iridium is shown in Fig. 1.

Although the location of vacancies is well-defined, the extent of the relaxation in the positions of the neighbouring atoms is usually difficult to assess. Qualitatively, however, the displacements around vacancies in close packed structures, such as the noble metals, are small, but larger relaxations may be expected around vacancies in the more open structures. In the case of interstitial atoms the situation is more complex as even the

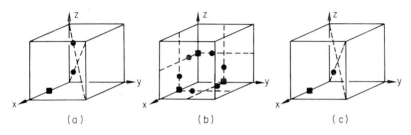

Fig. 2. Interstitial sites in (a) f.c.c. (b) b.c.c. and (c) diamond cubic structures. The sites shown are associated with only one lattice point of the cells. Thus, in f.c.c. crystals there is one octahedral site at $\frac{1}{2}[100]$ and two equivalent, slightly smaller sites at $\frac{1}{4}[111]$ and $\frac{1}{4}[113]$ associated with each atom. In b.c.c. crystals there are six equivalent sites at $\frac{1}{4}\langle012\rangle$ and a further three slightly smaller sites at $\frac{1}{2}\langle100\rangle$ for each atom. Finally, in the diamond structure there is one interstitial site for each atom of each of the two interpenetrating f.c.c. cells forming the structure; these are situated at $\frac{1}{2}[100]$ and $\frac{1}{4}[111]$.

locations of the intrusive atoms are often uncertain. It seems reasonable to assume however that they are situated in the largest holes of the structure. The positions of these holes in the f.c.c., b.c.c. and diamond cubic structures are indicated in Fig. 2. These holes are small for both the f.c.c. and b.c.c. structures, although there are three times as many available in the latter case. Small impurity atoms, such as carbon in iron, can therefore be accommodated at these positions without causing appreciable strains in the surrounding crystal, but larger atoms, such as displaced parent atoms, must of necessity, produce large displacements. In principle, this is not the case in the diamond structure as, on purely geometrical grounds, only one-half of the available lattice points are occupied. However, because of the special bonding characteristics associated with this structure, a large number of interstitials is not possible. Similarly, in the case of ionic crystals, the geometry of the lattice usually permits the presence of interstitials, but the difference in size of the different ions often means that the structure is virtually close packed, so that few interstitials are permitted. The strains around interstitial atoms in f.c.c. crystals are particularly large and suggest that in this case the associated relaxation may not be even

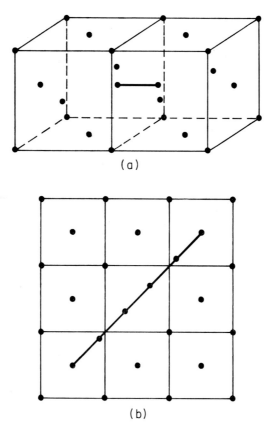

(a)

(b)

Fig. 3. Alternative configurations for interstitial atoms in f.c.c. crystals:
(a) the split interstitial, (b) the crowdion.

approximately spherically symmetric. Two alternative configurations, known as the split interstitial and the crowdion are shown in Fig. 3.

Individual point defects in ionic crystals are associated with an electric charge. Thus, for example, the region from which a positive ion is missing is associated with a net negative charge and *vice versa*. Therefore, in order that the crystal might be electrically neutral, such defects often occur in groups. In the simplest case, of crystals built up of two kinds of oppositely charged ions, the principal types are the Frenkel defect, consisting of a vacancy and an interstitial ion of the same sign, and the Schottky defect, consisting of a positive ion vacancy and a negative ion vacancy. The Anti-Schottky defect, which is a pair of oppositely charge interstitials, is also possible in principle but is thought to be rare in practice. The Schottky defect can occur in either the free form, in which the vacancies are separated or the associated form, in which they occupy neighbouring lattice points. These two situations, and also the Frenkel defect, are illustrated schematically in Fig. 4.

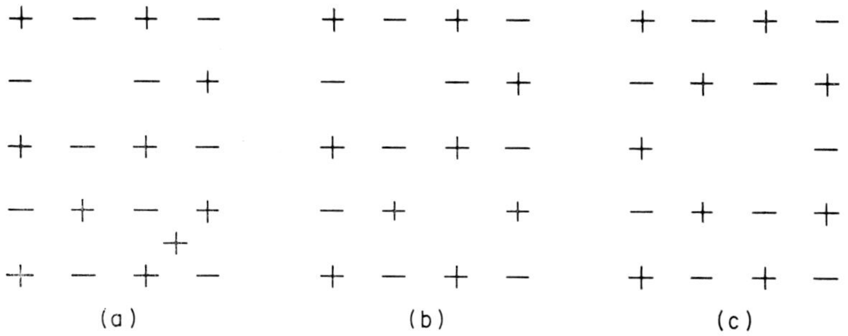

Fig. 4. Schematic representation of (a) Frenkel defect (b) Schottky defect and (c) Associated Schottky defect in a simple ionic crystal. No relaxations in the positions of neighbouring ions are shown.

Negative ion vacancies may also become elastically neutral by trapping an electron which is then shared by the neighbouring positive ions. Similarly, the net negative charge associated with a positive ion vacancy can be balanced by one of the neighbouring negative ions losing an electron. The vacancy is then said to be associated with a defect-electron or hole. The resulting imperfections affect the optical properties of the crystal and are thus known as colour centres, being labelled F-centres and V-centres respectively. Many other more complex point defects and colour centres also arise in ionic crystals, particularly when impurity ions of different charge are present.

The electrical conductivity of ionic crystals is primarily controlled by the diffusion of charged vacancies through the crystal, but the motion of free electrons and holes also makes a small contribution. In semiconductor materials it is this second term which is important. In pure semiconductors there is a high-temperature intrinsic conductivity, arising from holes and electrons being produced in equal numbers from broken bonds. Conduction at lower temperatures can only be achieved by introducing point defects. Thus, a substitutional trivalent atom, such as gallium, or a penta-valent atom such as antimony in silicon or germanium, produces holes and electrons respectively. These give rise to extrinsic conductivity. Vacancies and interstitial impurities which can be easily ionised will also give rise to this effect.

We have seen that the concentration and mobility of point defects in ionic crystals controls the electrical and optical properties of these materials.

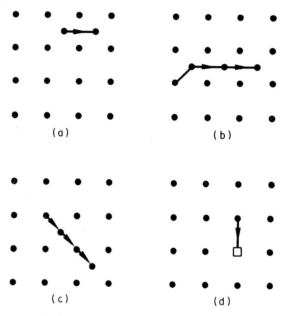

Fig. 5.   Possible diffusion mechanisms involving point defects: (a) interstitial diffusion of an interstitial atom, (b) interstitial diffusion, (c) interstitialcy diffusion, (d) vacancy diffusion.

In metals these quantities are also important as they are responsible for diffusion processes and hence may influence the mechanical properties. Some diffusion mechanisms involving point defects are illustrated in Fig. 5. In the noble metals Au, Ag and Cu, diffusion occurs primarily by the motion of vacancies. Therefore, diffusion is clearly enhanced by a large concentration of vacancies and high vacancy mobility. These quantities are fixed by the activation energies E_V^F and E_V^M to form and move a vacancy respectively, each of which is approximately 1 eV in the noble metals. The vacancy concentration c in thermal equilibrium at temperature T is given by $c = \exp(-E_V^F/kT)$ where k is Boltzmann's constant. The coefficient of self diffusion D depends on temperature in the same way, being given by $D = D_0 \exp(-Q/kT)$ where D_0 is independent of T and $Q = E_V^F + E_V^M$. Thus, both the concentration and the mobility of vacancies, and indeed of all point defects, are largest near the melting point, where the vacancy concentration may reach 10^{-4}. Large concentrations of defects can also be obtained at lower temperatures by rapidly quenching from near the melting point. Indeed, most experiments designed to provide information on point defects are carried out on either quenched or irradiated specimens.

The activation energies E_I^F and E_I^M to form and move interstitials in the noble metals are much larger and much smaller respectively than E_V^F and

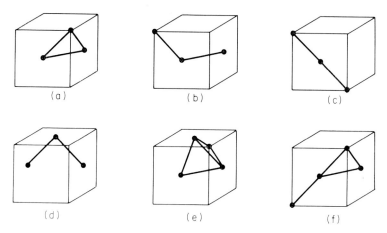

Fig. 6. Possible configurations for trivacancies (a–d) and tetravacancies (e and f) in f.c.c. crystals.

E_V^M. Thus, the very few interstitials that are present in thermal equilibrium are very mobile. Larger numbers of interstitials may be introduced as a result of radiation damage but, due to their high mobility, most of these are likely to be quickly annihilated at the crystal surface or internal sinks.

Another important type of point defect in the noble metals is the pair of associated vacancies known as the divacancy. The formation energy of this defect is rather smaller than $2E_V^F$ due to the presence of a binding energy, which may be of the order of 0·2 eV. Nevertheless, the concentration of divacancies in thermal equilibrium, which in this case is also influenced by the fact that this defect, unlike the vacancy, may be oriented in different crystallographic directions, will be much smaller than that of isolated vacancies. However, as the energy to move a divacancy is probably only about 2/3 eV, this defect may make a significant contribution to the diffusion processes. Larger groups of vacancies, such as trivacancies and tetravacancies, are probably only important when a crystal is supersaturated with defects, following rapid quenching or irradiation. Trivacancies in f.c.c. crystals may arise in four crystallographically distinct configurations, as shown in Fig. 6. Many more possibilities exist for the tetravacancy, two examples also being given in Fig. 6. These clusters probably act as nucleation sites for some of the large two- and three-dimensional defects described in later Sections.

3. LINE DEFECTS

Dislocations are the only important type of one-dimensional defect in crystals and, unfortunately, their characteristic features are comparatively difficult to describe. However, in order to obtain an adequate understanding of the ways in which dislocations influence the physical and mechanical properties of crystals, rigorous definitions of these features are necessary. The major part of this section thus deals with the basic concepts of dislocation geometry. This background material is then used to discuss a few special cases and some experimental observations of dislocations, but it also provides an introduction to the following section on two-dimensional defects.

It is convenient to introduce the idea of a dislocation by considering the processes by which slip occurs in crystals. The predicted stresses, necessary for the rigid body displacement of one part of a crystal over another, are much larger than those found in practice. This suggests that crystals deform by means of a process of local slip as shown in Fig. 7(a). In this mechanism the region of the crystal which has slipped, steadily increases in size until the whole crystal has sheared. At an intermediate stage, the boundary in the slip plane between the slipped and unslipped regions is then known as a dislocation line or simply a dislocation. The vector representing the magnitude of the local slip, or more generally the difference between the amounts of local slip in adjacent regions of the slip plane, is known as the Burgers vector of the dislocation and is normally represented by **b**. It is clearly associated with the whole dislocation and is thus a constant for the line. It follows that dislocations cannot end inside a crystal; they must either form closed loops, as in Fig. 7(a), or terminate at other dislocations, or the crystal surface. If the Burgers vector is a lattice vector, the crystal structure at the plane over which local slip has occurred remains unchanged, as shown in Fig. 7(*b*). Such a dislocation is truly a line defect and is termed perfect. Alternatively, if **b** is not a lattice vector, the resulting dislocation is associated with a sheet of defective crystal, as shown in Fig. 7(*c*). Imperfect dislocations of this type will therefore be discussed in the section on two-dimensional defects.

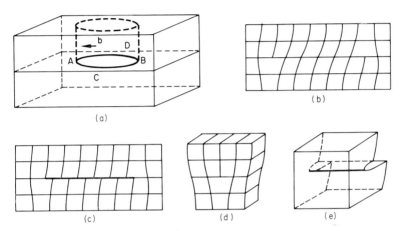

Fig. 7. Schematic representations of dislocations. In (*a*) a dislocation loop, shown in bold lines, is produced by local slip of magnitude and direction **b**, within the loop, of the upper part of the crystal. At A and B the dislocation has positive and negative edge character respectively and at C and D it has right-hand and left-hand screw character. Diagrams (*b*) and (*c*) illustrate the structure of the slipped region for the case of perfect and imperfect dislocations; the defective layer in (*c*) is shown by a bold line. The principal characteristics of a positive edge and a right-hand screw dislocation are indicated in (*d*) and (*e*).

The atomic configuration near a dislocation depends partly on the crystal structure, but also to a large extent, on the relative orientation of the Burgers vector and the dislocation line itself. Thus, when **b** is perpendicular to the line, as at A in Fig. 7(*a*), the strains are as shown in Fig. 7(*d*). This is an edge dislocation and it can clearly take up any orientation in a plane perpendicular to **b**. It is characterised by the extra half plane of atoms shown above the slip plane in Fig. 7(*d*). However, this half plane does not indicate that the edge dislocation is a sheet defect. The defective crystal is concentrated at the edge of this extra half plane so that the dislocation is truly a line defect. Note that the extra half plane of the dislocation at B in Fig. 7(*a*) is below the slip plane, as indicated in Fig. 7(*b*). The two cases are known as positive and negative edge dislocations respectively, but there is no real physical difference between them.

When **b** is parallel to the dislocation line, as at C in Fig. 7(*a*), the strain field is helical, as shown in Fig. 7(*e*). This is then a right-hand screw dislocation, the corresponding dislocation at D being a left-hand screw. In this case the two dislocations are physically distinct. Screw dislocations are, of necessity, straight lines parallel to **b**. In practice, most dislocations in crystals have both edge and screw components, and are termed mixed dislocations.

The definition of the Burgers vector of a dislocation in terms of the direction and magnitude of the local slip it produces on moving is, unfortunately, inadequate. For example the displacement shown in Fig. 7(*b*) could be produced either by shearing the upper half of the crystal to the left or the lower half to the right. This leads to ambiguity in the sign of the Burgers vector. A rigorous definition and sign convention are thus necessary. A widely used procedure is to place an arrow on the dislocation line and trace out a closed Burgers circuit, in a right-hand (RH) sense around the line. This circuit must be wholly contained in good-crystal, and avoid the bad-crystal near the dislocation line where the structure is unrecognisable. If the steps of the circuit are now repeated in a reference-crystal, that is a good-crystal from which elastic strains and thermal vibrations have been removed, the circuit will not close. The closure failure defines the magnitude of the Burgers vector, its sense being from the finishing point (F) to the starting point (S) of the circuit. This convention is conveniently denoted by the symbol (FS/RH). Examples of Burgers circuits in real and reference crystals for the cases of edge and screw dislocations are given in Figs. 8(*a*)–(*d*).

Using the (FS/RH) convention the displacement resulting from the motion of a dislocation can be established in the following way. Using the right hand, point the forefinger in the direction of the arrow on the dislocation line and the second finger in the direction of motion. The third finger then points to the side of the slip plane which is displaced in the direction of the Burgers vector. The application of this rule is illustrated in Fig. 8(*e*) and (*f*), for the case of a moving screw dislocation.

In discussing the behaviour of dislocations in crystalline materials, certain results of classical elasticity theory, applied to dislocations in infinite elastic media, are widely used. Although the emphasis of this chapter is on the structure of defects in crystals, it is convenient at this stage to summarise some of these results. In particular, the stress due to a dislocation is found to be inversely proportional to distance from the line, and the elastic energy to be approximately Gb^2, where G is the shear modulus. Near the dislocation line the elastic strains are large and do not obey Hooke's law; this region is known as the dislocation core and is usually ignored in elastic energy calculations. The force F acting on unit length of dislocation due to an applied stress p is pb. This force will tend to move dislocations through the medium but, if a dislocation is pinned at certain points along its length, it will bow out between these points until its radius of curvature R is given approximately by $R = Gb/p$.

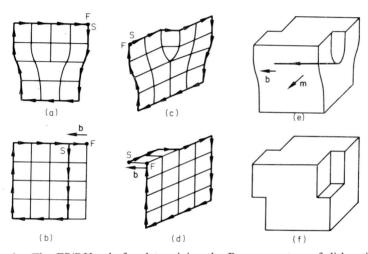

Fig. 8. The FS/RH rule for determining the Burgers vectors of dislocations and the displacements caused by their motion. In (a) and (c) Burgers circuits are drawn in good-crystal around edge and screw dislocations respectively. In (b) and (d) these circuits are repeated in reference-crystals. The start and finish of the circuits are denoted by S and F and the Burgers vector **b** is defined by FS as indicated. The displacement caused by moving the right-hand screw dislocation shown in (e) in the direction **m** is illustrated in (f).

Dislocations also exert forces on each other. Thus, like parallel screw dislocations repel, and unlike parallel screws attract each other. This is also true for parallel edge dislocations in the same slip plane. However, the situation is more complex when the edges lie in different glide planes and when the dislocations are not parallel. Elasticity theory can also be used to examine the interactions between dislocations and point defects. In particular it shows that, as might be expected, large solute atoms are attracted towards the region of extension associated with an edge dislocation, and small solutes to the region of compression.

The result of elasticity theory, that the energy of a dislocation is proportional to the square of the Burgers vector, is particularly useful in analysing dislocation interactions. For example, consider a perfect dislocation of Burgers vector 2**a**, where **a** is a primitive lattice vector. This dislocation can reduce its energy by a factor two, by dissociating into two perfect dislocations of Burgers vector **a**. It follows that the Burgers vectors of the dislocations which arise in practice, tend to be the shortest possible lattice vectors.

We shall now examine the motion of dislocations in more detail. Movement of a dislocation on the surface, defined by its own line and its Burgers vector, is known as glide. Edge and mixed dislocations are, therefore, restricted to glide on a definite surface, as envisaged in the discussion of local slip given above. The glide process involves only small movements of the atoms near the dislocation core, so that unpinned dislocations can move under the action of very small stresses. The minimum stress necessary to move a dislocation in an otherwise perfect crystal is known as the Peierls–Nabarro stress and can again be estimated using elasticity theory. As a perfect screw dislocation line is parallel to its own Burgers vector, it can, in principle, glide on any plane. The change from one glide plane to another is known as cross-slip.

An edge dislocation can only move out of its own slip plane by the addition or removal of material to or from the edge of its extra half plane. This process, which is known as climb, is much more difficult than glide. It is achieved in practice by point defects, usually vacancies, attaching themselves to, or detaching themselves from, the edge of the extra half plane. If the dislocation line is originally straight, the arrival of a vacancy results in two steps, known as jogs, being formed. The effective length, and hence the energy, of the line is thus increased, as shown at A in Fig. 9(*a*). However, if the vacancy attaches itself at a pre-existing jog, the dislocation configuration and energy remains unchanged, as shown at J in this diagram. Thus, the rate of climb is enhanced by the presence of large concentrations of vacancies and jogs.

Jogs may arise from the intersection of two dislocation lines. An example of this is illustrated in Fig. 9(*b*)–(*d*), where a gliding edge dislocation intersects a stationary screw. In Fig. 9(*b*) both dislocations are straight lines but, using the (FS/RH) rule described above, it may be seen that, following the intersection, both dislocations have acquired jogs, as shown in Fig. 9(*c*). The jog on the edge dislocation is a short segment of edge dislocation and can thus glide unhindered on the stepped slip surface. The jog on the screw dislocation is also a short segment of edge and is thus unable to glide on the slip surface shown in Fig. 9(*d*). In order to move with the rest of the dislocation line it must climb, and the (FS/RH) rule indicates that, for the situation shown in Fig. 9(*d*), this must be achieved by the emission, rather than by the collection, of vacancies.

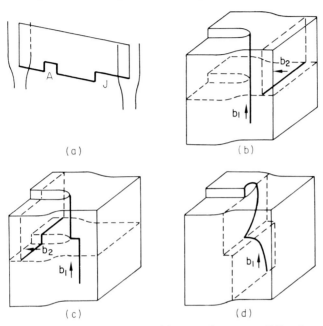

(a)   (b)   (c)   (d)

Fig. 9.   Jogs on dislocation lines. In (*a*) vacancies are annihilated at the edge of the extra half plane of an edge dislocation. In (*b*) and (*c*) jogs are produced when an edge dislocation of Burgers vector b_2 intersects a screw dislocation of Burgers vector b_1. The jog on the screw is unable to glide on the slip surface shown in (*d*).

However, this situation may not arise, because the jog on the screw dislocation of Fig. 9(*c*) may be eliminated by glide of the dislocation line or, equivalently, by the glissile motion of the jog along the dislocation.

The Burgers vector of the dislocation loop shown in Fig. 7(*a*) lies in the plane of the loop. This need not be the case. In particular, dislocation loops may be produced when vacancies or interstitials collect together to form a disc. This is illustrated for the case of an interstitial loop in Fig. 10(*a*). If the dislocation line forming the loop is perfect the loop can glide to take up any configuration on the cylinder, or prism, of axis **b** and bounded by the dislocation line, as shown in Fig. 10(*b*). The loop is then known as prismatic.

The formation of prismatic dislocation loops from vacancies is an effective way of increasing the density of dislocations in a crystal. This can also be achieved by vacancies diffusing to a jog on a screw dislocation line, which may result in the formation of a helical dislocation. This dislocation is produced by the systematic climb of the jog to produce a tightly wound helix of pitch equalling the Burgers vector. The configuration is energetically unfavourable however, due to the close approach of edge dislocation segments of the same sign. The loops of the helix therefore glide apart and sometimes long open helices are formed. Right- and left-handed screw dislocations do, of course, produce right- and left-handed helices respectively.

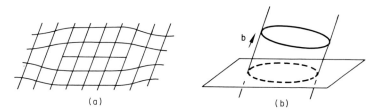

Fig. 10.   Dislocation loops. An interstitial loop is shown in (*a*) and the motion of a prismatic loop from its initial position (broken curve) to a new orientation (bold curve) in (*b*).

So far we have considered individual dislocation lines. In practice, dislocations in crystals are often linked to other dislocations to form elaborate three-dimensional networks. The fundamental element of such networks is the dislocation node, or triple-point, where three dislocations meet. The Burgers vectors of the three dislocations forming such a node must clearly be related in a special way. Using the concept of the Burgers circuit it is, in fact, found that, if the directional arrows on the dislocations all point towards the node, or all point away from it, the sum of the three Burgers vectors is zero. Beautiful two-dimensional networks of dislocations are also found, separating regions of crystal of slightly differing orientations. The limiting cases of these sub-grain boundaries are the twist-boundary and tilt-boundary, corresponding to orientations related by rotations about a line perpendicular to the boundary, and in the boundary respectively. These boundaries consist of networks of screw dislocations and a set of parallel edge dislocations respectively.

When a crystal is deformed the density of dislocations present increases rapidly. Thus, for example, an initial density of 10^6 cm of line per cubic cm

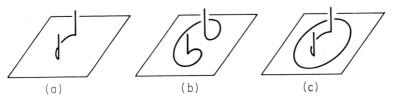

Fig. 11.   The operation of a Frank–Read source.

in a well-annealed metal, may increase to values exceeding 10^{12} cm^{-2}. A large number of multiplication mechanisms have been proposed but it now appears that most dislocations originate from grain-boundaries, crystal surfaces or precipitate particles. However, the best known and most elegant mechanism is the classic Frank–Read source, in which a segment of dislocation line bows out on its glide plane to form closed dislocation loops as shown in Fig. 11. Using the relation for the radius of curvature of a dislocation segment, which was quoted earlier, it is seen that such a source operates under a stress p given approximately by $p = 2Gb/l$, where l is the separation of the pinning points. Similar sources involving only one pinning point, and hence producing dislocation spirals, rather than loops, also occur. Analogous sources which operate by means of climb have also been reported.

The presentation of this section has so far been general and the illustrations schematic. The results can, in principle, be applied to all crystal structures, but in any given case the dislocations will have their own individual characteristics. For example, dislocations in many materials, including most metals, are closely associated with two-dimensional defects and are discussed in the next Section. Again, in ionic crystals the dislocation structure is restricted by the condition of charge neutrality. This is illustrated in Fig. 12(a) for the case of an edge dislocation in NaCl. The

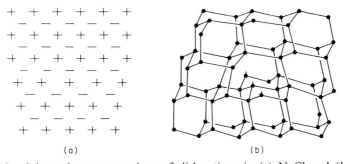

Fig. 12.   Schematic representations of dislocations in (a) NaCl and (b) Ge.

Fig 13(a)

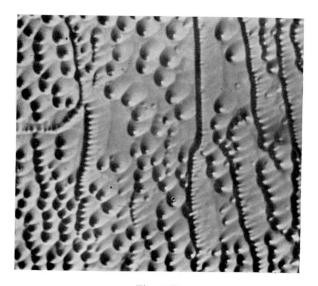

Fig. 13(b)

Fig. 13(c)

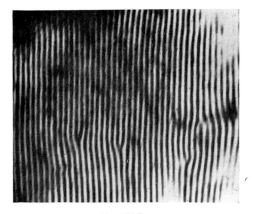

Fig. 13(d)

choice of an extra half plane is seen to be ambiguous but in each case equal numbers of positive and negative ions are involved. The ions at the edge of this half plane alternate in sign so that there is no net charge associated with a straight dislocation line. However, jogs on dislocations in ionic crystals are sometimes charged and can then interact with charged point defects. Materials with the diamond structure, such as Ge, are characterised by tetrahedrally disposed directed valence bonds between the constituent atoms. Along the axes of dislocations some of these bonds may be unsaturated, as shown in Fig. 12(b). The free or dangling bonds can trap electrons, and hence, greatly influence the electrical properties. Other

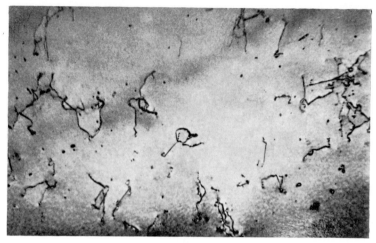

Fig. 13(e)

Fig. 13.   Optical micrographs (*a* and *b*) and transmission electron micrographs
(*c–e*) illustrating the direct observation of dislocations. (*a*) Trigons on an octahedral face of a diamond at the points where dislocations meet the surface
($\times$ 160). (*b*) Etch pits on the surface of a AgCl crystal indicating the sites of
dislocations some of which are aligned to form low angle boundaries ($\times$ 1400).
(*c*) A dislocation in planes of Pt atoms in a crystal of platinum phthalocyanine.
The extra half plane can be seen most clearly by looking tangentially along the
dark lines. The spacing of these lines is about 12 Å. (*d*) Dislocations in Moiré
patterns formed by superposition of {111} films of Au and Pd. The spacing of
the fringes is about 29 Å. (*e*) Dislocations and dislocation loops in Al strained
4% ($\times$ 22,000). Courtesy of K. E. Puttick, M. T. Sprackling, J. W. Menter,
D. W. Pashley and C. T. B. Foxon respectively.

interesting features arise in the case of materials with the zincblende
structure. For example, in InSb positive and negative edge dislocations
can be distinguished, as the edge of the extra half plane consists either of
all In atoms or of all Sb atoms. The two types of dislocation have different
effects on the electrical properties. Another characteristic of dislocations
in these materials is that the core energy depends critically on the orientation of the line. Therefore, the dislocations tend to lie along well-defined
crystallographic directions, as shown for the case of Si in Fig. 15. This
contrasts markedly with the behaviour of most dislocations in metals, as
shown for example in Fig. 13(*e*).

Much of the basic dislocation theory described in this Section was
formulated before the existence of dislocations was confirmed experimentally. However, in recent years, a vast amount of direct experimental
information on defects has been obtained and the micrographs of Fig. 13

illustrate the results of some of the different experimental techniques that have been used on a variety of materials. The earliest methods involve surface observations, the classic example being spiral growth features associated with the emergence of dislocations with a screw component. Of more general application is the study of the small pits which may occur at the points where dislocations meet the surface of crystals. These pits may have a marked crystallographic character, as in the case of the trigons on a octahedral plane of a diamond shown in Fig. 13(*a*), or have the mammillary form exhibited by the etch pits on AgCl, shown in Fig. 13(*b*). However, it is the observation of dislocations, using transmission electron microscopy techniques, with resolution better than 10 Å, that has provided the greatest wealth of experimental information. The classic experiment in this case was the direct resolution of planes of Pt atoms, of spacing about 12 Å in crystals of platinum phthalocyanine. The characteristic extra half plane of edge dislocations could therefore be observed, as illustrated in Fig. 13(*c*). Unfortunately, because of the limited resolution, this method can only be used for a few carefully chosen materials. However, when two thin crystals of the same structure but of different lattice parameters are superposed in the same orientation, a system of relatively widely spaced Moiré fringes is produced with the electron microscope, which clearly indicates the presence of dislocations in the crystals. An example of such a pattern, arising from superposed {111} layers of Au and Pd is shown in Fig. 13(*d*). Finally, Fig. 13(*e*) illustrates the most widely used transmission electron microscopy technique, in which the dislocations are made visible through the diffraction of electrons by the dislocation strain fields. In this example, of a foil of Al strained 4%, an irregular arrangement of dislocation lines and a number of small dislocation loops can be seen. Further examples of electron micrographs, illustrating two-dimensional defects are given in the next section.

4.   SHEET DEFECTS

Much of the discussion in the last section was general, in that it could be applied to perfect dislocations in all materials, regardless of crystal structure. In the present section a great deal of this generality will be lost, as two-dimensional defects are, of necessity, crystallographic in character. This also applies to the imperfect dislocations, bounding sheets of defective crystal, which will also be examined here.

The principal forms of two-dimensional defects are faults in the stacking sequences of certain crystallographic planes. The simplest, and those which have received the greatest amount of attention, are faults in f.c.c. and h.c.p. crystals. In these structures the faults lie on the octahedral {111} planes and the basal (0001) planes respectively. In perfect crystals these planes are stacked—ABCABCA—and—ABABABAB—respectively and faults may be introduced in a variety of ways. Thus, the f.c.c. sequence may be reversed at any stage to give a twinned structure—ABCA*CBA—for example, in which the right and left hand parts of the crystal are related by reflection in the twin boundary A*. It is impossible to form such a fault on the basal planes of the h.c.p. structure. However, two other simple faults on the f.c.c. octahedral planes may be produced by either removing a plane or inserting a plane, to give the sequences—ABCBCAB—and —ABCB*ABC—respectively. It will be noted that, in the latter case, the plane introduced, B*, is labelled differently from the adjacent planes; this ensures that the nearest neighbour distances in the crystal are preserved. The fault obtained by removing a plane may also be produced by displacing part of a perfect crystal by $\frac{1}{6}\langle 112 \rangle$ so that, for example, A → B, B → C, C → A. If the nearest neighbour rule is applied to h.c.p. structures, the only analogous fault is that obtained by inserting a plane C* to give —ABABC*ABA—. However, if in addition to removing or inserting a plane, the part of the crystal on one side of the fault is displaced by $\frac{1}{3}\langle 1\bar{1}00 \rangle$, so that, for example, A → B, B → C, the two faults obtained are identical, as indicated below:

$$-\text{ABABABAB}- \rightarrow -\text{ABAB BABA}- \rightarrow -\text{ABABCBCB}-$$

$$-\text{ABABABAB}- \rightarrow -\text{ABABCABA}- \rightarrow -\text{ABABCBCB}-$$

If an offset of this kind is applied without inserting or removing material, one obtains the faulted stacking sequence—ABABCACA—. All six faults, and also the perfect stacking sequences, are plotted schematically in Fig. 14.

The characteristic features of the faults described above may be deduced by examining the stacking sequences in detail. In fact, although the nearest neighbour distances are preserved, the faults differ considerably in the next nearest neighbour violations. Thus, the f.c.c. twin has one violation, labelled CAC, of the perfect ABC sequence, whereas, the other two f.c.c. faults each have two violations of the same type. These are labelled BCB and CBC for the fault obtained by removing a plane, and BCB and BAB for that obtained by inserting a plane. The two faults are not equivalent, however, as the defective crystal extends over a wider band of material in the latter case, as clearly shown in Figs. 13(c) and (d). These two faults do, in fact, characterise two fundamentally different types of stacking fault, known as intrinsic and extrinsic respectively; these may be present in any crystal structure. An intrinsic fault is one in which the perfect stacking sequence, on either side of the fault, extends up to the composition plane itself, no material being excluded; an extrinsic fault is one which does not

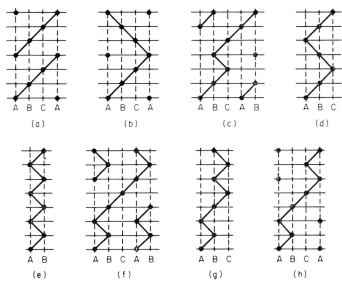

A B C A   A B C A   A B C A B   A B C
  (a)   (b)   (c)   (d)

A B   A B C A B   A B C   A B C A
 (e)   (f)   (g)   (h)

Fig. 14. Possible faults in f.c.c. and h.c.p. structures: (a) f.c.c. perfect, (b) f.c.c. twin, (c) f.c.c. intrinsic, (d) f.c.c. extrinsic, (e) h.c.p. perfect, (f) h.c.p. extrinsic, (g) h.c.p. single intrinsic, (h) h.c.p. double intrinsic.

have this property. Considering now the faults in the h.c.p. structure shown in Figs. 14(f)–(h), one finds *three* next nearest neighbour violations of the type ABC in the fault obtained by simply inserting a plane, *one* if there is an additional offset and *two* if the offset occurs alone. The first fault is extrinsic and the other two are both intrinsic.

The principles used above to examine possible faults in the f.c.c. and h.c.p. structures are quite general and can thus be applied to any crystal structure. Thus, for example, it is suggested that faults on the {112} planes of b.c.c. crystals, which have a six-fold stacking sequence, may be represented by—ABCDEFE*F*ABCDEF—. This fault, which is intrinsic in character and involves two nearest neighbour violations, labelled EFE and FEF, may be obtained by inserting the two planes E* and F* or by displacing part of the crystal by $\frac{1}{6}\langle 111 \rangle$. It is important in the theory of deformation twinning of b.c.c. crystals. Note that for planes with a stacking sequence greater than three-fold, it is not possible to produce faults with no nearest neighbour violations by simply inserting or removing a single plane.

Stacking faults can be observed experimentally using either electron diffraction or X-ray diffraction techniques. In both cases the faults produce a characteristic fringe pattern. This is illustrated in Fig. 15, which shows a

Fig. 15.   X-ray diffraction topograph of a stacking fault and several dislocations in Si ($\times$ 36). Courtesy of A. E. Jenkinson.

large fault on a $\{111\}$ plane in Si, together with several dislocation lines. This micrograph was taken using the X-ray procedure, which suffers from the disadvantage of being unable to achieve high magnification. Useful information can only be obtained from crystals with a low defect density but, as this method can be used on bulk specimens, it provides valuable direct information which is otherwise unobtainable. Stacking faults can also be seen on the electron micrographs of Figs. 18, 19 and 22.

Faults in the stacking sequence of a set of crystal planes are associated with an additional contribution to the energy of the crystal. At a first approximation the energy may be assumed to be proportional to the number of next nearest neighbour violations. This implies that, in the case of f.c.c. crystals, the stacking fault energies of intrinsic and extrinsic faults are approximately equal, and twice that of the twin boundary. However, in practice, the extrinsic fault, because it extends over more planes, is usually thought to have a larger energy than the intrinsic. In h.c.p. crystals it is likely that the fault obtained by inserting a plane is always accompanied by an offset, thus reducing the effective energy by a factor three. The stacking fault energy is a characteristic feature of a given crystal structure, and is a major factor in determining its physical and mechanical properties.

The preceding discussion tacitly assumed that stacking faults extend throughout a crystal. In practice, a fault may terminate within a crystal the boundary then being a line of imperfect dislocation. The Burgers

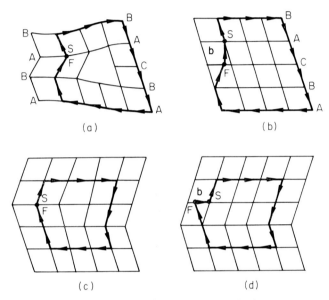

Fig. 16. Burgers circuits for imperfect dislocations in f.c.c. structures. In (a) and (c) circuits are drawn around a Frank partial dislocation and a twinning dislocation respectively. These are repeated in the reference-crystal in (b) and (d) enabling the Burgers vectors **b** to be specified.

vector of such a dislocation will be given by the displacement which produces the fault and, may or may not, lie in the plane of the fault. The corresponding types of dislocation are known as Shockley and Frank partials respectively. The sense of the Burgers vector is given by a Burgers circuit which, in order to be performed in good crystal, must start and finish at the fault. An example of a circuit of this kind is illustrated in Fig. 16(a) and (b), for the Frank partial bounding the fault formed by removing an octahedral plane in a f.c.c. crystal. It defines a Burgers vector given by $\frac{1}{3}\langle 111 \rangle$. The Frank partial obtained by inserting an octahedral plane has the same Burgers vector but, as it is attached to a different fault, is structurally different. In the h.c.p. structures Frank partials may have Burgers vectors given by $\frac{1}{2}[0001]$, corresponding to simply inserting a basal plane, or by $\frac{1}{2}[0001] + \frac{1}{3}\langle 10\bar{1}0 \rangle = \frac{1}{6}\langle 20\bar{2}3 \rangle$, corresponding to insertion or removal of a plane, plus offset. Shockley partial dislocations are produced by slip and for f.c.c. and h.c.p. crystals, corresponding to the faults shown in Fig. 14(c) and (h), have Burgers vectors given by $\frac{1}{6}\langle 112 \rangle$ and $\frac{1}{3}\langle 10\bar{1}0 \rangle$ respectively. As imperfect dislocations must remain attached to their associated stacking faults and, like perfect dislocations, can only move conservatively in the plane defined by their line and Burgers vector, it follows that Shockley partials are glissile, their motion extending or contracting the area of the fault, but Frank partials are sessile, being unable to move except by a climb process.

Unlike the other stacking faults a twin boundary cannot terminate at a simple dislocation line. However, it may move from one plane to a neighbouring parallel plane, as shown in Fig. 16(c) and (d) to give rise to a line defect known as a twinning dislocation. In this case it is not possible to perform a conventional Burgers circuit around the dislocation line, but by making a slight generalisation of this concept, as shown in Fig. 16(c) and (d), a unique Burgers vector may be specified. In particular, in f.c.c. crystals the Burgers vector of the twinning dislocation is the same as that of the Shockley partial.

In practice, the removal or insertion of a crystal plane may be achieved by the condensation of vacancies or interstitials to form a dislocation loop. If the resulting dislocation is perfect, the loop is termed prismatic, as described in Section 3, but in many cases faulted loops bounded by Frank partial dislocations are produced. Such loops may be transformed into prismatic loops, however, if a Shockley partial dislocation sweeps across the loop. Thus, if a Frank loop forms on a (111) plane of a f.c.c. crystal, it will have a $\frac{1}{3}[111]$ Burgers vector, which can be changed to a perfect $\frac{1}{2}[110]$ slip dislocation if a $\frac{1}{6}[11\bar{2}]$ Shockley partial annihilates the fault, as shown in Fig. 17. In this way the energy of the dislocation line itself, which is proportional to the square of its Burgers vector, is increased by a half, but the energy of the system as a whole may be reduced, due to the removal of the stacking fault.

In some cases doubly faulted loops, consisting of a triangular loop within a hexagonal loop, may be produced in f.c.c. crystals. In such a case two planes have been removed from the region within the triangle but only one plane from the region between the triangle and hexagon. The corresponding faults are extrinsic and intrinsic respectively, and the

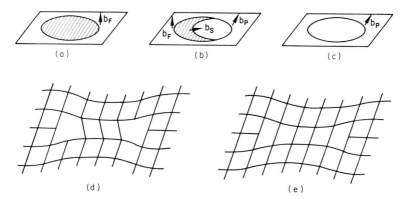

Fig. 17. Removal of the stacking fault of a Frank dislocation loop. The original faulted loop (*a*) has Burgers vector b_F. The fault is removed (*b*) by a Shockley partial dislocation of Burgers vector b_S sweeping across the loop, thus producing the prismatic loop (*c*) of Burgers vector b_P. The initial and final structures of the loop are shown in (*d*) and (*e*).

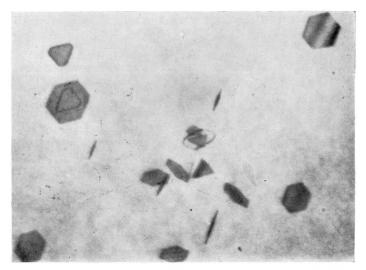

Fig. 18. Electron micrograph of dislocation loops in quenched high purity Al. The loops lie on the four variants of the {111} planes, one of which is approximately perpendicular to the foil. Examples of single faulted hexagonal loops and two types of double faulted loops, consisting of faulted triangles within either faulted or unfaulted hexagons, are present (× 32,000). Courtesy of J. W. Edington and R. E. Smallman.

annealing characteristics of the loops can, in fact, be used to determine values of the stacking fault energies. The characteristic shapes of these loops are believed to be determined by the anisotropy of the dislocation core energies. Some examples of different kinds of dislocation loops in quenched Al are shown in the electron micrograph (Fig. 18).

The dislocation reaction which transforms Frank sessile loops into glissile prismatic loops may be reversed so that a perfect dislocation with Burgers vector $\frac{1}{2}[110]$ may split into a Frank partial of Burgers vector $\frac{1}{3}[111]$ and a Shockley partial of Burgers vector $\frac{1}{6}[11\bar{2}]$. This dissociation mechanism indicates the origin of the term partial dislocation and is

Fig. 19. Electron micrograph of extended dislocations, showing stacking fault contrast, in Cu–12 at.% Al alloy strained 6 % ($\times$ 40,000). Courtesy of J. A. M. Salter.

known to occur in semiconductor crystals with the diamond structure. However, a much more important mechanism is the dissociation of perfect slip dislocations into pairs of Shockley partials joined by ribbons of stacking fault. Thus, for example, in f.c.c. crystals a perfect dislocation of Burgers vector $\frac{1}{2}[1\bar{1}0]$ lying in the (111) plane can dissociate into two Shockley partials with Burgers vectors $\frac{1}{6}[2\bar{1}\bar{1}]$ and $\frac{1}{6}[1\bar{2}1]$. This results in a reduction of the elastic self energy of the dislocations by a factor two-thirds. This implies that the two Shockley partials repel each other. They therefore separate, creating a ribbon of stacking fault, which in turn increases the energy of the system until an equilibrium separation is reached. An example of this is shown in Fig. 19, which is an electron micrograph of extended dislocations in a Cu–Al alloy of low stacking fault energy. Note the marked difference between this micrograph and Fig. 13(*e*) which shows dislocations in Al, which has a high stacking fault energy. In hexagonal materials, perfect dislocations with Burgers vectors equalling $\frac{1}{3}[11\bar{2}0]$ dissociate into Shockley partials with Burgers vectors $\frac{1}{3}[10\bar{1}0]$ and $\frac{1}{3}[01\bar{1}0]$. Similar reactions occur in other crystal systems.

When dislocations dissociate in the way described above, the directions in which the two Shockley partials move are fixed by the crystal structure. In order to fully specify these directions, and indeed to study dislocation interactions in general, it is convenient to introduce a notation for slip

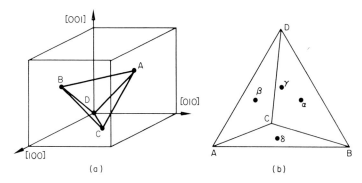

Fig. 20. The Thompson tetrahedron.

planes and Burgers vectors which does not use Miller indices. The best known and most widely used notation of this type is that of the Thompson tetrahedron. This is used for f.c.c. crystals, but similar geometrical models have been introduced for other crystal systems. The Thompson tetrahedron is formed from the four {111} planes, as shown in Fig. 20(*a*), the edges being the six $\frac{1}{2}\langle 110\rangle$ vectors. Its corners are labelled A, B, C, D and the centres of the opposite faces α, β, γ, δ, as indicated in Fig. 20(*b*). The Burgers vectors of $\frac{1}{2}\langle 110\rangle$ perfect dislocations may then be represented by pairs of Roman letters, such as **AB**, the order of the letters indicating the sense of the dislocation. Similarly $\frac{1}{3}\langle 111\rangle$ Frank partial dislocations are given by a Roman letter and the corresponding Greek letter, such as **Aα**, and $\frac{1}{6}\langle 112\rangle$ Shockley partials by the remaining combinations of Roman and Greek letters, such as **Aβ**. The two dissociations discussed above can then be written **AB → Aα + αB** and **AB → Aγ + γB**. The rule for deciding the sense of the dissociation is now as follows. Looking from outside the tetrahedron in the direction of the arrow on the dislocation, the partial represented by letters in the order **R**oman–Greek is on the **R**ight.

An important illustrative example of the application of the Thompson notation arises in the analysis of dislocation networks. In Fig. 21(a), a

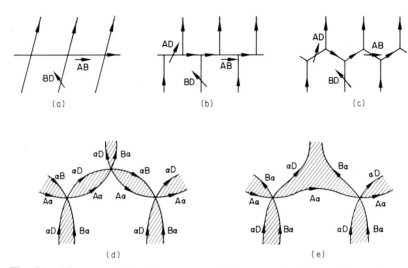

Fig. 21.   The use of the Thompson notation in analysing the formation of extended and contracted nodes. Areas of stacking fault are shown shaded.

screw dislocation with Burgers vector **AB** intersects a set of parallel mixed dislocations **BD** lying in the plane γ. The dislocations interact elastically to form short segments of dislocation of type **AD**, as shown in Fig. 21(b). All three dislocations and Burgers vectors lie in γ so that, due to the line tensions acting at the triple points, the equilibrium configuration shown in Fig. 21(c) is produced. If the dislocation segments between the nodes now dissociate the situation shown in Fig. 21(d) arises. At half of the nodes, adjacent Shockley partials are now identical and may therefore link up; this will not be possible at the remaining nodes. Thus, the network of extended and contracted nodes shown in Fig. 21(e) is finally obtained. Similar reactions can occur in other crystal systems, and the extended nodes have received a great deal of attention, as they enable experimental values of the stacking fault energy to be determined. An example of a network of screw dislocations forming a twist boundary on the basal plane of graphite, which has a very small stacking fault energy and thus exhibits

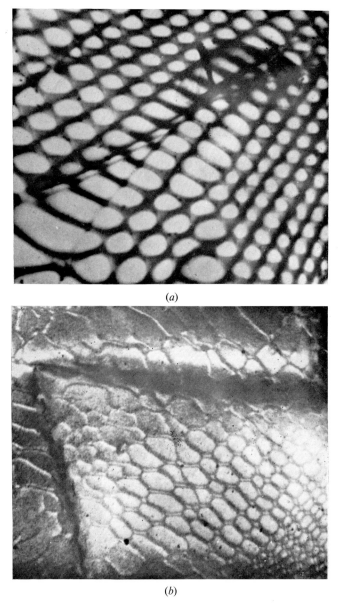

(a)

(b)

Fig. 22.   Electron micrographs showing networks of dislocations in (a) graphite
($\times$ 21,000) and (b) polyethylene ($\times$ 20,000). Courtesy of J. A. Hedley and
V. F. Holland respectively.

extended nodes very clearly, is shown in Fig. 22(*a*). The accompanying micrograph Fig. 22(*b*) shows a more complicated network lying on the (001) plane of crystalline polyethylene, which has an orthorhombic structure. In this case, the three dislocations meeting at a node are still essentially screw in character, but are not crystallographically equivalent; the Burgers vectors are [110], [100] and [010], the [110] dislocations being widely extended.

Perfect screw dislocations can, in principle, slip on any plane but, in practice, they are, like edge dislocations, restricted to move on specific crystallographic planes. However, in f.c.c. crystals dislocations of this kind, with $\frac{1}{2}\langle 110 \rangle$ Burgers vectors, lie in two $\{111\}$ slip planes and may

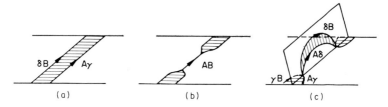

Fig. 23.   Cross slip of an extended screw dislocation (*a*), which becomes constricted (*b*) before moving on to a different slip plane (*c*). Areas of stacking fault are shown shaded.

therefore cross-slip between these planes. For example a screw dislocation **AB** in Fig. 20(*b*) may cross-slip between planes γ and δ. This process is particularly important in theories of hardening, as it enables dislocations to glide past barriers. Perfect dislocations may of course be dissociated into $\frac{1}{6}\langle 112 \rangle$ Shockley partials, which lie in only one $\{111\}$ plane, thus making cross slip difficult. However, if the stacking fault energy is high, and the distance of separation of the partials therefore small, an applied stress can force the partials together, thus enabling cross slip to occur. This is illustrated in Fig. 23 where a dislocation of Burgers vector **AB**, dissociated into **A**γ + γ**B** on plane γ, has been constricted and then dissociated into **A**δ + δ**B** on plane δ.

The type of dislocation with Burgers vector given by pairs of Greek letters in the Thompson notation, has not yet been considered. These dislocations occur in practice when dissociated dislocations on intersecting slip planes interact to form a wedge of stacking fault. The simplest and most important case is when dislocations **BC** and **CA**, which are dissociated as **B**α + α**C** and **C**β + β**A** on planes α and β respectively, interact to give **B**α + (α**C** + **C**β) + β**A**, as shown in Fig. 24. The dislocation α**C** + **C**β = $\alpha\beta$

is then known as a stair-rod, as it keeps the carpet of stacking fault fixed at the line of intersection of the two planes. Like the Frank partial it is sessile, its Burgers vector being of type $\frac{1}{6}\langle 110 \rangle$. Other stair-rod dislocations, at both acute and obtuse stacking-fault bends, may arise but the one described involves the greatest reduction in elastic energy, two Shockley partials, each with energy proportional to 1/6, combining to form a dislocation with energy proportional to 1/18. This effectively means that the interaction is very likely to occur in practice, and that, once the stair-rod is formed, it is very difficult to dissociate into its component Shockley partials. Thus, being sessile, it forms an effective barrier to the

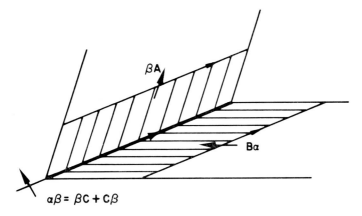

Fig. 24.   A stair-rod dislocation at the line of intersection (bold) of two slip planes. The associated stacking faults are shown shaded.

movement of other dislocations. The whole configuration of stacking faults and stair-rod dislocation is, in fact, known as a Lomer–Cottrell lock.

In ordered alloys, the passage of a dislocation with the normal Burgers vector may destroy the order across the slip plane, thus producing a two-dimensional defect known as an antiphase boundary. However, if a second dislocation, with the same Burgers vector, moves across the same slip plane, the order is restored, as illustrated schematically in Fig. 25, for an ordered two-dimensional crystal. Therefore, in these materials, dislocations tend to be linked by means of a ribbon of antiphase boundary and to move in pairs. The distance of separation of the component dislocations will, of course, depend on the energy of the antiphase boundary. In practice, this energy is small, so wide ribbons may be observed. Moreover, each perfect dislocation, comprising such super-lattice dislocations,

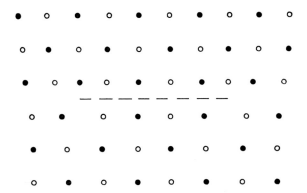

Fig. 25.   Schematic representation of a super dislocation consisting of a ribbon of antiphase boundary in an ordered alloy.

can itself dissociate into Shockley partials so that fourfold ribbons may be produced. Since the ordering of these alloys may commence at different nuclei within a given grain, antiphase boundaries may be produced where these separate domains meet. Indeed, periodic arrays of antiphase boundaries may develop, as illustrated in the electron micrograph Fig. 26.

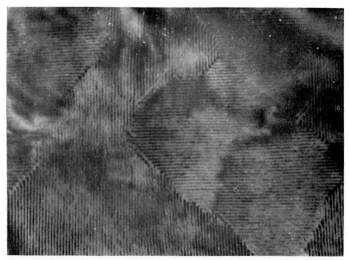

Fig. 26.   Electron micrograph of ordered CuAu II which consists of periodic arrays of antiphase boundaries. The spacing of the boundaries is about 20 Å. Courtesy of D. W. Pashley.

5. VOLUME DEFECTS

In Section 2 we discussed point defects occurring either individually or in small groups. We shall now briefly consider some structural imperfections consisting of aggregates of large numbers of these defects. Clusters of vacancies, interstitials and solute atoms often occur in crystals and may condense to form volume defects such as voids and precipitates with a variety of different morphological features controlled by the structure. For example cubic cavities may be produced in impure KCl crystals by annealing in hydrogen. The cavities grow by means of vacancy diffusion and may be many microns across. Vacancies arriving at a cavity form growth nuclei, and hence, steps on the internal surfaces, which act as preferential sites for further vacancy condensation. Voids are also observed in quenched Al crystals, but in this case are octahedral in shape, being surrounded by {111} planes. These defects may play an important intermediate stage in the formation of dislocation loops.

Clusters of defects are produced when crystals are irradiated and may condense to form volume defects. One case of particular interest, is the formation of bubbles containing one of the inert gases. The solubility of these gases in metals is very low, but they may be introduced by bombarding a foil with their energetic ions using a particle accelerator. Alternatively, they may be produced by transmutation during neutron irradiation of reactor materials, such as those containing U and Be. The gas precipitates to form bubbles which coarsen on heating and may produce an appreciable volume increase. This swelling is technologically important as it can set a limit on the use of important reactor materials. Bubbles of helium gas in Cu foils bombarded with α-particles have been studied in detail. It is found that the pressure in the bubbles is balanced, at least approximately, by the surface energy, and is thus inversely proportional to the radius of the bubbles. The bubbles may migrate in the foils by a process of diffusion of atoms over their surfaces. They may also coalesce and of course, when they reach foil surfaces, explode.

In Au, and some other f.c.c. metals, vacancy condensation may result in volume defects known as stacking fault tetrahedra. These consist of four equilateral triangles of intrinsic stacking fault, forming the faces of a regular tetrahedron. Two different mechanisms have been proposed for the formation of these defects and both are thought to operate in practice. The first involves the continuous growth of the tetrahedron through the absorption of vacancies. The smallest defect of this kind which can be envisaged arises from the trivacancy configuration shown in Fig. 6(a).

It occurs if the atom, forming the apex of the regular tetrahedron which has the triangle of three vacancies as base, is displaced to the centre of this tetrahedron. As the tetrahedron grows its faces will, in general, contain steps which act as preferred sites for the condensation of additional vacancies. In the second mechanism the stacking fault tetrahedron is produced from a faulted dislocation loop lying in a $\{111\}$ plane. The Frank partial dislocations surrounding the loop dissociate into stair-rod dislocations lying along the three $\langle 110 \rangle$ directions in the plane of the loop and Shockley partial dislocations on the other three $\{111\}$ planes. The stair-rod dislocations form the base of the tetrahedron and the Shockley partials glide on their respective slip planes and interact to form further stair-rod dislocations along the other three $\langle 110 \rangle$ directions, which meet at the apex of the tetrahedron. The Thompson notation, discussed in Section 4, is a valuable tool in describing the detailed crystallography of these defects.

From a technological standpoint, the most important volume defects are precipitates, which may be classified into three structural types, according to the nature of their interfaces. In coherent precipitates a number of parent atoms in a localised region are replaced by an equal number of solute atoms. As the sizes of the two species of atoms will not be the same these precipitates are associated with long-range elastic strain fields. Secondly, incoherent precipitates are formed by a given volume of matrix being replaced by an equal volume of second phase. Although such defects have no associated elastic strain fields their incoherent interfaces do, in general, have high energies. Finally, semi-coherent precipitates may arise in which the interfaces consist of regular networks of dislocation lines. In such a case the elastic strain energy and interfacial energy are both comparatively small. An illustrative example of the formation of coherent and semi-coherent precipitates is provided by the age-hardening Al alloys. In these, coherent Cu-rich platelets about 50 atoms in diameter and one atom thick, may form on $\{100\}$ planes as a result of low temperature ageing. The relaxation of the crystal around these platelets is similar to that around a platelet of vacancies which, as discussed in Section 3, gives rise to a prismatic dislocation loop. In the present case the effective Burgers vector is much smaller than that of the loop, perhaps being only about one-sixth of the $\{100\}$ interplanar spacing. Thus, if a stack of six of these coherent zones can form on adjacent planes, the combined elastic strain field can be accommodated by the formation of a dislocation around the perimeter of the precipitate, which thus becomes semi-coherent. The platelet is only one of the morphological forms which precipitates may assume, spherical and polyhedral types also being common. However regardless of type, shape and size, all precipitates act as barriers to the easy movement of dislocations in crystals and thus, when present, control the mechanical properties.

6. CONCLUSIONS

In this Chapter an attempt has been made to introduce the reader to the different types of structural imperfections which can exist in all crystalline materials. For some defects, particularly dislocations, a thorough under-standing of their basic geometrical characteristics is necessary before their structural configurations in specific crystals can be considered. In such cases superficial treatments are clearly undesirable and comparatively detailed general accounts have been given. Applications to particular structures and materials have therefore had to be limited, and those examples which have been treated in detail, have been primarily concerned with the structure of defects in crystals of high symmetry. These are clearly the simplest cases to consider, but the approach adopted is quite general and can be applied, with only minor modifications, to all crystals. The analysis is, however, naturally more involved, and the possible defects more difficult to portray diagrammatically, for crystals of low symmetry. Nevertheless, as the range of materials of commercial importance is extended, the study of the fine structure of defects in specific crystals is becoming one of the most significant aspects of defect theories.

Brief references have also been made in this chapter, mainly through the medium of micrographs illustrating the text, to the variety of different experimental techniques used to observe defects. The main application of these techniques is not, of course, merely the observation of defects, but the investigation of the influence of these defects on the physical and mechanical properties of crystals. An indication has been given here of how defects control the plastic deformation of crystals and, in some cases, the electrical properties. They also play an important role in determining many of the magnetic, superconducting, optical and thermal properties of crystalline materials. The possible fields of application for the theory of defects in crystals are thus extensive. Some of these are discussed in detail in succeeding chapters of this volume; other works dealing with this topic are listed under the heading Further Reading.

7. FURTHER READING

Specific textual references have been deliberately avoided in this chapter. More detailed information may be obtained from the following books, which have comprehensive bibliographies of original research papers. References 1 and 10 are particularly useful as they deal with defects in a variety of crystalline materials, 7 and 11 provide excellent elementary introductions to dislocation theory and 6 and 8 are more advanced recent accounts of the subject. Applications of defect theory in the metallurgical field are reviewed in 2 and 3; 4, 5 and 9 are valuable earlier works.

1. Amelinckx, S. (1964). *The Direct Observation of Dislocations*, Academic Press, New York.
2. Cahn, R. W. (Ed.) (1965). *Physical Metallurgy*, North-Holland, Amsterdam.
3. Christian, J. W. (1965). *The Theory of Transformations in Metals and Alloys*, Pergamon, Oxford.
4. Cottrell, A. H. (1953). *Dislocations and Plastic Flow in Crystals,* Oxford University Press, London.
5. Friedel, J. (1964). *Dislocations*, Pergamon, Oxford.
6. Hirth, J. P. and Lothe, J. (1968). *Theory of Dislocations*, McGraw-Hill, New York.
7. Hull, D. (1965). *Introduction to Dislocations*, Pergamon Press, Oxford.
8. Nabarro, F. R. N. (1967). *Theory of Crystal Dislocations*, Oxford University Press, Oxford.
9. Read, W. T. (1953). *Dislocations in Crystals*, McGraw-Hill, New York.
10. van Bueren, H. G. (1961). *Imperfections in Crystals*, North-Holland, Amsterdam.
11. Weertman, J. and Weertman, J. R. (1964). *Elementary Dislocation Theory*, Macmillan, New York.

CHAPTER 3

MECHANICAL TWINNING

D. HULL AND E. ROBERTS

1. INTRODUCTION

Friedel[1] defined a twin as 'a polycrystalline edifice built up of two or more homogeneous portions of the same crystal species in juxtaposition and oriented with respect to each other according to well-defined laws'. This definition applies to all forms of twin, namely, *growth* twins which form during solidification by growth at the solid–liquid interface, *annealing* twins which form during recrystallisation of a cold-worked material, and *mechanical* twins which form by a shear process in response to an applied stress. In general, growth and annealing twins exert only a secondary influence on the deformation behaviour of solids, whereas mechanical twinning is an important mechanism of plastic deformation in a large number of crystalline materials.

The major part of this chapter is devoted to crystallographic and geometrical aspects of mechanical twinning, since this is most appropriate in relation to the topics covered by other authors. By limiting the subject in this way it is possible to give an effective introduction to the understanding of the role of mechanical twinning in a wide variety of problems, such as the relation between mechanical twinning and fracture, and the role of twinning as a lattice invariant deformation process during martensitic transformations. This type of problem is beyond the scope of this chapter. Most of the work reported has been done in the past fifteen years. Detailed accounts of the subject before this period have been given by Cahn[2] and Hall.[3] No comparable reviews have been written in recent years although Christian[4] has an excellent chapter in his book, and much useful information is given in the book by Klassen-Neklyudova.[5] A conference on deformation twinning was held in Gainesville in 1963 and the proceedings have been published (see Reed-Hill, Hirth and Rogers[6]).

2. FORMAL CRYSTALLOGRAPHY OF TWINNING

The earlier work on mechanical twinning by such workers as Johnsen[7] and Niggli[8] assumed that during the twinning transformation a cell of the parent crystal sheared directly to a cell in the twin, the twin and parent lattices being identical. Implicit in this assumption is the idea that the twin lattice orientation is a mirror image of the matrix lattice across the composition plane, or alternatively, that the twin lattice orientation is obtained from the matrix orientation by a rotation of 180° about the direction of the twinning shear lying in the composition plane. These ideas are still valid for most twins, although recent work by Crocker[9] and Bevis and Crocker[10] has shown that other twins not satisfying this simple relationship are possible in some crystal structures. For this reason the definition of a mechanical twin, formulated by Bilby and Crocker, is preferred, thus, 'a mechanical twin is a region of a crystalline body, which has undergone a homogeneous shape deformation (pure shear), in such a way that the product structure is identical with that of the parent, but oriented differently'.

The rudiments of the formal crystallography of mechanical twinning are well documented (*see*, for example, Schmid and Boas,[11] Cahn[2], and Hall[3]), but is included here to introduce the terminology used subsequently. The *elements* of twinning can be defined by reference to a sphere representing the matrix crystal, which is sheared as illustrated in Fig. 1. The plane K_1 is called the *habit* or *composition* plane and η_1 is the direction in this plane parallel to the shear displacement. When half of the sphere is sheared parallel to K_1 in the η_1 direction to form a twin the new shape of the hemisphere is half of a triaxial ellipsoid with semi-axes a, b, and c. The diametrical plane forms the interface between the twin and parent crystal. During shear all directions contained in planes parallel to the interface remain invariant. Thus the dimensions and shape of the K_1 plane remain unaltered and for this reason K_1 is called the *first undistorted plane*. This plane is truly invariant, since not only does its shape remain unaltered, but the orientation is unchanged also. Further examination reveals that after shear there is a second plane which maintains its shape and has the same dimensions as K_1. This plane K_2 is called the *second undistorted plane*, but it is not truly invariant because its orientation is changed by the shear. Figure 1 shows that the plane K_2 intersects the first undistorted plane K_1 along the axis a of the triaxial ellipsoid which lies normal to the direction η_1; this axis has a magnitude equal to the radius of the original sphere. The plane containing the direction η_1 and perpendicular to K_1 is called the *plane of shear*, S. The line of intersection of S and K_2 is labelled η_2, and K_2 and η_2 shear to K_2' and η_2' respectively. The plane K_2 subtends an angle 2φ with the first undistorted plane K_1, this is the angle between the directions η_1 and η_2. The orientation of K_2 with respect to K_1 is

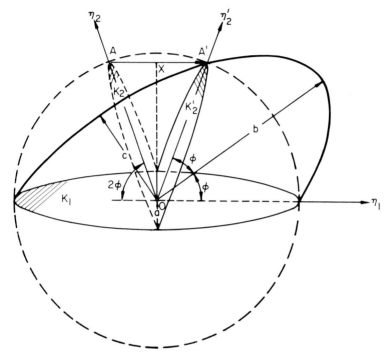

Fig. 1.   The shear of a sphere into an ellipsoid illustrating the twin elements which define a given twin mode.

directly related to the magnitude of shear $s = AA'/OX$ which can be determined from the twinning elements. From Fig. 1 s is given as

$$s = 2 \cos 2\varphi \qquad (1)$$

The line which bisects the directions η_1 and η_2' is the semi-major axis of the ellipsoid b and in the same plane normal to b is the semi-minor axis c. The planes K_1, K_2 and directions η_1, η_2 are called collectively the *twinning elements*. Although a twinning shear is completely defined by K_1 and η_2 or K_2 and η_1, it is customary to quote all four twinning elements together with the magnitude of the shear strain, s.

The classical treatments of deformation twinning impose the restriction that one lattice cell of the parent structure must shear to become an equivalent cell of the twin. The resultant twinning modes may be grouped conveniently into two classes, and following Cahn[2] these two classes of twinning modes are referred to as *type I* and *type II twins*.

2.1. Twins of type I

With this twin K_1 and η_2 are rational, that is the elements K_1 and η_2 have rational indices, and, in general, K_2 and η_1 are irrational. The fundamental changes which occur during type I twinning are illustrated in Fig. 2. The directions x_1 and x_2 are any two rational directions in the composition plane, and η_2 is the only rational direction in the irrational plane K_2. These three directions may define a primitive or multiply primitive cell of the Bravais space lattice of the unsheared crystal. After shear of K_2 and η_2 to K_2' and η_2' of the twin, the directions η_2, x_1 and x_2 remain unaltered in length as well as in relative orientation with each other. In addition, the angles between the directions x_1 and η_1' and x_2 and η_2' are

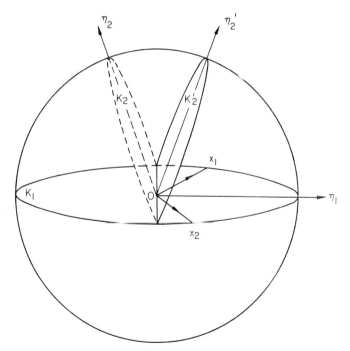

Fig. 2.   Illustration of parameters used to define type I twinning.

complementary to their original values. Thus, the three directions x_1, x_2 and η_2' in the sheared crystal, define a cell of the Bravais lattice which is identical to a cell of the unsheared lattice. The sheared part of the crystal is termed a *twin of type I*. The orientation of the twin lattice with respect to that of the matrix can be described in two ways:

(*a*) reflection through K_1

(*b*) rotation of π about the normal to K_1.

For the Bravais space lattice these two relationships are equivalent, but when crystal structure is considered, for example, when there is more than one atom per lattice site, this is not necessarily true. It should be noted that because of the relationships between the twin and the parent matrix it follows, first, that K_1 cannot be a symmetry plane of the lattice, and secondly, that the normal to K_1 cannot be an axis of two-fold symmetry.

2.2. Twins of type II

With this twin K_2 and η_1 are rational. Suppose that x_3 and x_4, Fig. 3, represent two rational directions in K_2 and η_1 represents the only rational

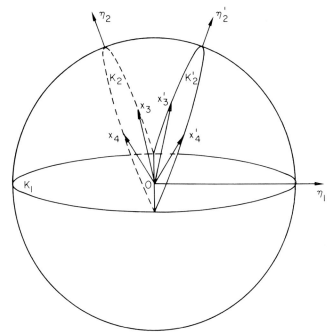

Fig. 3. Illustration of parameters used to define type II twinning.

direction in K_1. Following the discussion for type I twins, the directions x_3, x_4 and η_1 define a primitive or multiply primitive cell of the lattice. After the shear of K_2 to K_2' the angles between x_3' and x_4' remain unaltered and the angles between x_3', x_4' and η_1 are complementary to the original. Again, the directions x_3', x_4' and η_1, define a cell of sheared lattice identical with that of the original. The sheared crystal is termed a *twin of type II* and the orientation between twin and matrix can be described by:

(a) rotation of π about η_1

(b) reflection through the plane normal to η_1.

These two relationships are equivalent in terms of the Bravais space lattice, but may differ when account is taken of the crystal structure. Because of these relationships η_1 cannot be a two-fold axis within the crystal and the plane normal to η_1 cannot be a symmetry plane.

By considering only the orientation relationships associated with type I and type II twinning, discussed above, without making the assumption that a cell of the parent is sheared into an equivalent cell of the twin, Bilby and Crocker[12] deduced that the twinning elements K_1, η_2 and K_2, η_1 must be rational for type I and type II twinning respectively.

It is clear from Figs. 2 and 3 that for every type I twinning mode, defined by K_1 and η_2, there exists a type II twinning mode having the same magnitude of shear and with twinning elements $\bar{K}_1, \bar{K}_2, \bar{\eta}_1, \bar{\eta}_2$ given by $\bar{K}_1 = K_2, \bar{K}_2 = K_1, \bar{\eta}_1 = \eta_2$ and $\bar{\eta}_2 = \eta_1$. Such a mode is said to be the *conjugate* or *reciprocal* twinning mode to the first mode.

Twinning modes may have more than the two rational twinning elements $K_1, \bar{\eta}_2$, and K_2, η_1 associated with type I and type II twinning modes respectively. A *compound* twin (Cahn[13]) has four rational twinning elements. For twins of this form the orientation relationship may either be of type I or type II. However, when the plane of shear S is a symmetry plane these orientation relationships are equivalent. If a twin is compound then its reciprocal mode must be compound also. Crocker[14] has pointed out that further degenerate twin modes, with three rational elements, arise. Thus, some type I twinning modes may have K_1, η_2 and η_1 or K_1, η_2 and K_2 as rational elements with K_2 or η_1 irrational respectively. Similarly, a type II twin mode may have just K_1 or η_2 as the irrational twinning element.

Types of twin which do not satisfy any of the simple geometrical relationships described above have been reported, for example, by Crocker,[9] with respect to double shear processes in hexagonal metals, and, more generally, by Bevis and Crocker.[10] For these additional orientation relationships all four twinning elements may be rational, just K_1 and η_1 or K_2 and η_2 may be irrational[15] or in some cases the twin may be compound.[16] Considerable progress has been made in solving the formal crystallography of such twins,[10,16] but a great deal remains unknown and the subject will not be elaborated further here.

3. TRANSFORMATION OF CRYSTALLOGRAPHIC INDICES

This subject has been dealt with in detail by previous authors (*cf.* Hall[3] and Pabst[17]) and will be considered only briefly here. Suppose that two points A and B, in Fig. 4, in a Bravais space lattice define directions parallel to η_2 of a type I twin (in general, AOB is not in a straight line), and lie equidistant on each side of a potential composition plane K_1. If the co-ordinates of A and B with respect to the primitive reference axes of the Bravais space lattice are x, y, z and x', y', z' respectively, then

$$U:V:W = (x - x'):(y - y'):(z - z')$$

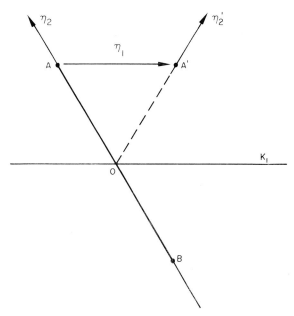

Fig. 4.   Transformation of crystallographic indices by a twinning shear.

and

$$H(x + x') + K(y + y') + L(z + z') = 0$$

where (HKL) and $[UVW]$ define the rational twinning elements K_1 and η_2. These two equations yield the following solutions for x', y', z',

$$x' = x - 2U\delta$$

$$y' = y - 2V\delta$$

$$z' = z - 2W\delta \qquad (2)$$

where $\qquad \delta = \dfrac{Hx + Ky + Lz}{HU + KV + LW}$

After formation of the twin the point A lies at A', Fig. 4, where it is a reflection of the point B through the composition plane K_1. Since the plane K_1 is invariant and point A' is a mirror reflection of B, the co-ordinates of A' with respect to the *new* reference axes of the Bravais space lattice, may be regarded as being numerically equivalent to those of B, namely, x', y' and z' represent the transformation of the co-ordinates of A during twinning.

If the direction $[uvw]$ is parallel to the line passing through both the co-ordinates x, y, z and the origin of the reference axes, then

$$u:v:w = x:y:z$$

After twinning this direction is transformed to $[u'v'w']$ where

$$u':v':w' = x':y':z'$$

Hence the transformation of direction indices is given by the equations

$$u' = u - 2U\beta$$

$$v' = v - 2V\beta$$

$$w' = w - 2W\beta \qquad (3)$$

where $\qquad \beta = \dfrac{Hu + Kv + Lw}{HU + KV + LW}$

The transformation of indices of a plane (hkl) within the primitive lattice, defined by the co-ordinates $(0, 0, 0)$ (x_1, y_1, z_1) and (x_2, y_2, z_2), can be determined in the same way. If the new indices of the plane are $(h'k'l')$ solution of the above equations gives

$$
\left.
\begin{aligned}
h' &= h - 2H\alpha \\
k' &= k - 2K\alpha \\
l' &= l - 2L\alpha \\
\end{aligned}
\right\} \tag{4}
$$

where
$$
\alpha = \frac{Uh + Vk + Wl}{HU + KV + LW}
$$

These formulae apply to any crystal systems and are used as the basis of an analysis on the transformation of dislocation vectors in Section 10. They also apply equally to the transformation and crystallographic indices during type II twinning. This can be verified by letting two directions $(u_1 v_1 w_1)$ and $(u_2 v_2 w_2)$ define the second undistorted plane K_2 such that

$$
u_1:v_1:w_1 = (x_1 - x_1'):(y_1 - y_1'):(z_1 - z_1')
$$
and
$$
u_2:v_2:w_2 = (x_2 - x_2'):(y_2 - y_2'):(z_1 - z_2')
$$

where x_1', y' ... etc. define both the co-ordinates of a point which is equidistant with the point $x_1 y_1$ ... etc. from the composition plane K_1 and the transformed co-ordinates of $x_1 y_1$ ... etc. with respect to the new reference axes. The indices (HKL) and $[UVW]$ in the formulae now refer to the rational twinning elements K_2 and η_1 respectively. However, it should be noted that the transformation formulae derived above apply only to orientation relation (a) for type I twinning and orientation (b) for type II twinning, and must not be confused with the transformation formulae used for determining which planes and directions in twinned crystals are parallel to planes and directions in the parent crystal (cf. Andrews and Johnson[18]). These formulae have been used extensively in the calculation of electron diffraction patterns from twinned crystals and the relationship between the two types of transformation, which are closely related to the twinning shear itself, have been discussed by Bevis.[19] Using the notation given above, a direction $[u'v'w']$ in the twin crystal is related to a parallel direction $[uvw]$ in the parent crystal by eqn. (3) where the indices $[UVW]$ are now the indices of the direction normal to the type I (orientation relation (a)) composition plane (HKL). Likewise a plane $(h'k'l')$ in the twin is parallel to the plane (hkl) in the twin when the indices are related by eqn. (4), $[UVW]$ again being the indices of the direction normal to (*HKL*). These formulae also apply to type II (orientation relation (b)) twinning when [*UVW*] are the indices of η_1 and (*HKL*) are the indices of the plane normal to η_1.

4. CONSERVATIVE INTERSECTION OF TWIN LAYERS

The term *twin intersection* was first used by Cahn[13] to describe the situation where one twin passes right through another, with or without change of direction, rather than tapering to a point and starting afresh on the other side, or producing extensive detwinning in the region of the interaction. The use of intersection in this narrow sense is rather confusing and

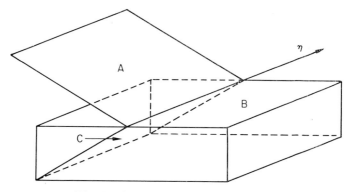

Fig. 5.   Conservative twin intersection.

for this reason such an intersection will be referred to here as a *conservative intersection*. The theory of this type of intersection was proposed by Cahn[13] for specific examples in α-uranium. In general, two or more twins can undergo a conservative intersection, provided that the directions of shear are in the same sense and are of identical magnitudes. If the composition planes of the twins are not parallel to one another then the directions of shear must lie along the only direction common to both. A specific example is illustrated in Fig. 5. The twin *B* forms first and is crossed by a second twin; *A* and *C* represent the composition planes of the *crossing* and *secondary* twins respectively. *A* and *C* represent a conservative intersection with the crossed twin *B*, as a result of co-operative shears along the common direction parallel to their line of contact, which is also contained in the composition plane K_{1B} of the crossed twin.

Two forms of conservative intersection can occur depending on the type of the crossed twin B. If this is a type I twin then the composition planes of twins A and C will be of the same index with respect to their parent lattices, since they form mutual reflections of each other through the composition plane K_{1B}. The requirement for a conservative intersection is thus reduced to that of finding two twins with respect to a single parent lattice which meet each other along a direction parallel to the shear direction η_1 of one of them. If the crosssed twin is type II the composition planes of A and C will be related to one another through a rotation of π about η_1 of the crossed twin. Cahn[13] has shown that for this intersection the two twins resulting in the intersection cannot have the same indices with respect to their parent lattices. Since the sense and magnitude of the twinning shears must be identical and lie along the direction common to both twins the requirement for a conservative intersection reduces to finding two different twins, such as (hkl) and $(h\bar{k}l)$, of the same family, or two conjugate twins, which have the same shear but belong to different families, which satisfy the condition for conservative shear. Liu[20] has shown that there is a second type of conservative twin intersection, based on the same general principles, outlined above, in which the crossing twin remains straight and the crossed twin is bent.

Another manifestation of twin intersections occurs in some crystal structures where two or more twins of the same family utilise the same

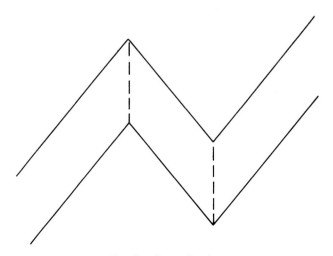

Fig. 6. Crossed twins.

direction and magnitude of shear. For these twins the resulting orientations of the twin lattices are identical, so that, when the twins meet, the
conditions for conservative shear will be automatically satisfied. This type
of intersection is illustrated in Fig. 6 and is usually called *crossed twins*
because of the analogy they make with the cross-slip process in which
simultaneous conservative glide occurs on two non-parallel planes along a
common glide direction.

5. DETERMINATION OF THE TWINNING ELEMENTS

To determine the twinning elements it is necessary to determine only K_1
and η_2 for type I twins, and K_2 and η_1 for type II twins. From a knowledge
of these the remaining elements and the magnitude of shear can be
deduced. The identification of type I twins is easier than type II since, in all
experimental analyses, the composition plane K_1 is the first to be determined. The overriding problem is whether or not the orientation of the
twin as well as the orientation of the matrix can be determined. In many
structures the twins are too fine to be examined with conventional X-ray
techniques, although micro-focus X-ray methods, transmission electron
microscopy and the recently developed back reflection electron beam
microprobe technique (Bevis and Swindells[21]) have removed the difficulty
to some extent.

5.1. Matrix lattice oriented

The simplest method of determining K_1 involves an examination of the
twin on two surfaces at a known angle. The orientation of the parent
lattice is determined using diffraction techniques and the position of the
traces of the twin matrix interface on the two surfaces is measured. The
crystallographic indices of K_1 can then be determined providing the twin
boundary is reasonably coherent. If the trace of the composition plane
cannot be followed on two surfaces the slope of this plane can be obtained
by successive grinding of the reference surface and measuring both the
new position of the twin and the depth of grinding.

The shear direction η_1 can be obtained if the trace of K_1 can be followed
along two inclined surfaces using the method of Greninger and Troiano.[22]
The two surfaces are prepared to an optically smooth finish prior to
twinning. The twinning shear produces a tilting of the surface through an
angle which depends on the orientation of these surfaces and the sense and
direction of η_1. A number of methods can be used to measure the angle

of the tilts and from these η_1 is determined. One method is illustrated in Fig. 7. The surface tilts produced by the twin are shown schematically in Fig. 7(a) and the poles of the surfaces before and after tilting are plotted on the stereographic projection in Fig. 7(b). These poles have been rotated in Fig. 7(c) so that the plane of the stereographic projection is parallel to K_1. Point a in Fig. 7(c) represents the line of intersection AB (Fig. 7(a)) of planes S and T and similarly point b the line of intersection BC of planes Sa and Ta. The change in orientation from a to b defines the effect of the twinning shear on this line originally oriented along a. The plane containing these two directions must contain the direction of shear η_1, which, by definition, must also be contained in the K_1 plane. Thus η_1 can be found directly by a simple construction. The sense of the shear is obtained unambiguously from Fig. 7(c) since a shears to b and not *vice versa*.

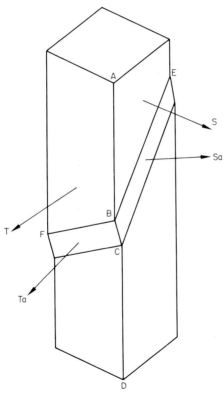

Fig. 7(a)

The two great circles defining the second undistorted plane K_2 and K_2' can be constructed by a process of trial and error since these two planes subtend equal angles with the plane perpendicular to η_1, as well as intersecting the planes S and Sa, and T and Ta along directions (c and d, and e and f, Fig. 7(c)) which lie symmetrically about this normal plane. The poles of K_1 and K_2 and the directions η_1 and η_2 lie on the common plane of shear s. The direction η_2 lies at the intersection of plane S and the second undistorted plane K_2. Finally, the angle φ between η_1 and η_2 can be measured and the twinning shear s calculated.

For twins in which it is not possible to measure the tilt angle it is necessary to make a detailed examination of the structure to obtain the remaining twinning elements after K_1 has been determined. If the crystal possesses a plane of symmetry which is normal to K_1 then this plane

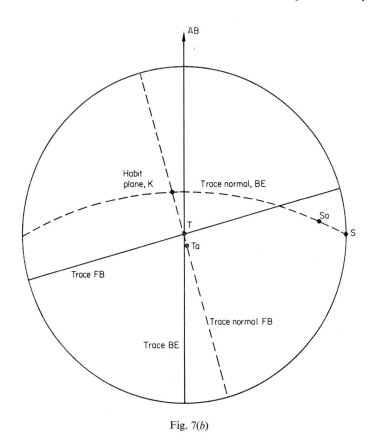

Fig. 7(b)

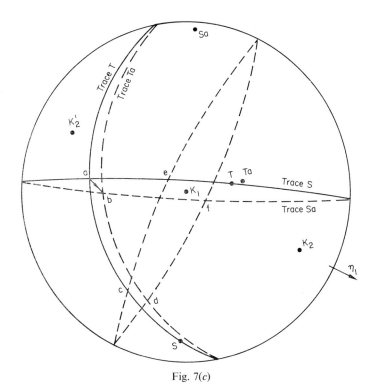

Fig. 7(c)

Fig. 7. Experimental determination of the twinning elements using the method developed by Greninger and Troiano[22]. (a) Tilts produced on two faces of a single crystal by the twinning shear. (b) Stereographic projection showing the orientation of the crystal surfaces before and after twinning. (c) As for (b), after the poles have been rotated to bring the plane of projection parallel to K_1, illustrating the construction required to determine the shear direction, η_1.

must correspond to the plane of shear S and two conclusions may be made:

(a) If K_1 is rational then η_1 must be rational also, and the twin must be either a degenerate type I compound twin with K_1, η_2 and η_1 rational, or a true compound twin with all four elements rational.

(b) If K_1 is irrational the twin must be a degenerate type II compound twin with K_2, η_1 and η_2 rational.

For *type I twins* with K_1 rational, a rational direction η_2 lying in the plane of shear S and close to the pole of K_1 must be sought. In general, the angle between the pole of K_1 and η_2 is not likely to exceed 25°, since this will involve unnecessarily large shears ($s > 0.93$). If several potential directions are found it should be possible to eliminate all but the correct one. The most probable direction will be the one which is closest to the

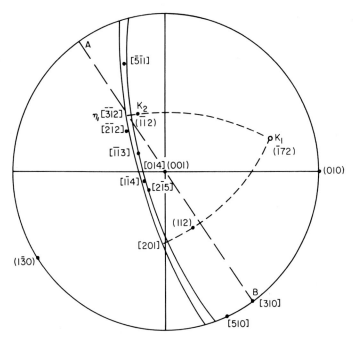

Fig. 8.   Determination of η_1 of (172) twins in α-uranium (after Cahn).[13]

pole of K_1, although it should be noted that, in general, the preferred twin will be determined by minimising both the *shuffles* and the shear, see Section 7. Since K_1 cannot be an axis of two-fold symmetry the choice will be reduced to two possibilities one on each side of the pole of K_1. With this situation it only remains to determine the sense of the shear, which can be obtained, if suitable reference lines are inscribed on the crystal surface before twinning.

For *type II twins* the first task is to determine the shear direction η_1. This can be obtained graphically, using the stereographic projection, since η_1 is defined by the intersection of K_1 and S. Since the pole of K_1 is only measured approximately, the plane being irrational, the location of η_1, which is rational, can now be used to establish K_1 more accurately.

The pole of the second undistorted plane K_2 is unlikely to be more than about $25°$ from η_1 in the plane S. Since this plane contains a rational direction η_2, which must correspondingly lie close to the pole of K_1 the complete determination of the twinning elements can be obtained as before; if the choice reduces to two possibilities, the correct identification can be made after determining the sense of the twinning shear.

If the crystal does not contain a plane of symmetry perpendicular to K_1 the determination of the twinning elements is more complicated but using a process of trial and error, the same type of approach can be applied with reasonable success. An example of an irrational type II twin in α-uranium is illustrated in Fig. 8 (Cahn[13]). The composition plane K_1 was found to lie close to (172). Because of the uncertainty of its exact position the two planes of K_1 were drawn to take account of the possible experimental errors. Between these two possible planes potential directions for the rational η_1 direction were plotted. Possible poles for the K_2 plane, having an angle of not more than about $20°$ from η_1, were also plotted on the same projection, such that the pole of K_1 and the proposed pole of K_2 lay on a great circle containing the potential shear direction η_1. From the diagram it can be seen that the most probable indices of K_2 and η_1 reduce to $(\bar{1}\bar{1}2)$ and $[\bar{3}\bar{1}2]$, and (112) and [201]. Of these two possibilities the first would produce a shear $s = 0{\cdot}228$ and the second a shear $s = 0{\cdot}702$. Thus, the most probable twinning elements were identified as $K_2 = (\bar{1}\bar{1}2)$ and $\eta_1 = [\bar{3}\bar{1}2]$. These elements were established subsequently by Cahn using additional experimental information.

5.2. Parent and twin lattice oriented

If the orientation of both the twin and the matrix can be determined experimentally the procedures for determining the twinning elements unambiguously are much simpler. The twins need to be examined on one surface only, and no assumptions concerning the coherency of the twin-matrix interface are required. The twinning elements can be deduced by superimposing a stereographic projection of the two lattices with respect to the surface of observation.

Type I—The composition plane, which is rational, will be parallel to two planes of like index in the projections of both lattices. If more than two such planes exist some can be eliminated by superimposing on the projection the trace normal of the twin composition plane, since this must contain the pole of K_1. In addition, coincident planes, which are symmetry planes, can also be eliminated. From any remaining possibilities K_1 can be determined unambiguously, since all like planes in both the twin and parent lattices, will be mutual reflections of each other through K_1. The remaining elements and the twinning shear can be determined using the methods already described.

Type II—The shear direction which is rational will coincide with a direction of like index contained in both the twin and parent lattices. If several such directions are common to the twin and parent lattices little elimination will be possible by a cursory examination with the exception of all directions corresponding to axes of twofold symmetry. The task of identifying η_1 is simplified by limiting the possible directions to those which lie within 25° of poles of like rational planes contained in both lattices. Further limitations are imposed by the requirement that all rational planes in the parent lattice must coincide with those in the twin lattice when they are rotated 180° about η_1. When η_1 has been identified the other elements can be obtained as before.

6.   STRAIN IN TWINNING

Formulae for the strain induced by twinning are derived in many texts, *e.g.* Schmid and Boas,[11] Hall[3] and Klassen-Neklyudova.[5] The calculations will be outlined briefly. Consider any point x, y, z with respect to three orthonormal axes X, Y and Z, Fig. 9. These axes can be selected in any desired manner but for convenience they have been chosen so that K_1 is normal to the Z-axis and η_1 parallel to the Y-axis. The twinning shear displaces the point x, y, z to a point with coordinates $x, y + zs, z$ where s is the twinning shear. Taking l_0 and l_1 as the magnitudes of the vectors from the origin to the points x, y, z and $x, y + zs, z$ respectively the twinning strain is

$$\varepsilon = \frac{l_1}{l_0} - 1$$

where $l_0 = (x^2 + y^2 + z^2)^{\frac{1}{2}}$

and $l_1 = (x_2 + y^2 + s^2 z^2 + 2szy + z^2)^{\frac{1}{2}} = (l_0{}^2 + s^2 z^2 + 2szy)^{\frac{1}{2}}$

If the initial vector subtends a maximum angle χ with the K_1 and an angle λ with η_1 then

$$z = l_0 \sin \chi$$

and

$$y = l_0 \cos \lambda$$

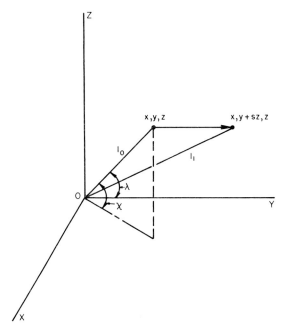

Fig. 9.   Calculation of the strain associated with a twinning shear.

Therefore,

$$l_1 = l_0(1 + 2s \sin \chi \cos \lambda + s^2 \sin^2 \chi)^{\frac{1}{2}}$$

and the twinning strain is

$$\varepsilon = (1 + 2s \sin \chi \cos \lambda + s^2 \sin^2 \chi)^{\frac{1}{2}} - 1 \qquad (5)$$

For a given value of λ, the maximum strain is obtained for $\lambda = \chi$. Substituting in the above expression the maximum strain can be calculated since $\partial\varepsilon/\partial\lambda = 0$ for maxima or minima. This gives

$$\lambda = \tan^{-1} \left(\frac{s \pm (s^2 + 4)^{\frac{1}{2}}}{2} \right)$$

which gives a maximum strain

$$\varepsilon = \pm \tan \lambda - 1 \qquad (6)$$

During twinning extension and contraction occurs simultaneously along different directions. The maximum *extension* is

$$\varepsilon = \frac{s + (s^2 + 4)^{\frac{1}{2}}}{2} - 1 \qquad (7)$$

and the maximum *contraction* is

$$\varepsilon = \frac{-s + (s^2 + 4)^{\frac{1}{2}}}{2} - 1 \tag{8}$$

Referring to Fig. 1 the final direction of maximum extension lies along the semi-major axis of the ellipsoid b and the final direction of maximum contraction lies along the semi-minor axis c. The ratios of the lengths of the three ellipsoidal axes are

$$a{:}b{:}c = 1 : \frac{s + (s^2 + 4)^{\frac{1}{2}}}{2} : \frac{-s + (s^2 + 4)^{\frac{1}{2}}}{2} \tag{9}$$

A knowledge of the strain associated with a particular twin is important in determining whether or not a twin will form in response to an applied stress or, alternatively, an externally imposed strain. When the elements of a twin mode are known the twin likely to form for a particular direction of strain can be illustrated on a stereographic projection. Thus, if the *whole crystal* is converted to a twin, all directions contained between the acutely

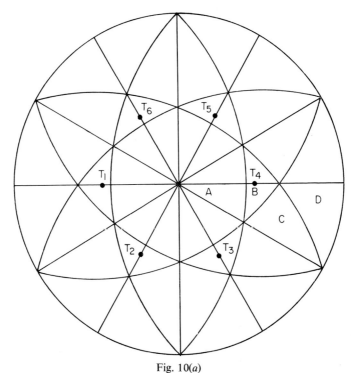

Fig. 10(*a*)

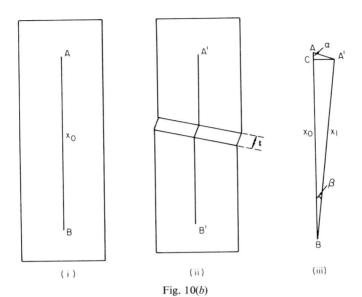

Fig. 10(*b*)

Fig. 10. (*a*) Change in length as a function of orientation for {10$\bar{1}$2} twins in zinc, consequent on complete twinning. T_1–T_6 are poles of twinning planes. *A* 1–6 contraction. *B* 2, 3, 5, 6 contraction; 1, 4 extension. *C* 2, 5 contraction; 1, 3, 4, 6 extension. *D* 1–6 extension (after Frank and Thompson[23]). (*b*) Calculation of the change in length due to a thin twin.

inclined planes K_1 and K_2 will shorten, and all directions contained in the obtuse angle will lengthen. An example is illustrated in Fig. 10(*a*) for twins in zinc which is self-explanatory. It follows that if a tensile stress was applied along any direction in region *A* of Fig. 10(*a*) none of the twins would form because all directions in this region are contracted during twinning. Similarly, if a compressive stress was applied along a direction in *C*, twins 2 and 5 would be favoured, and twins 1, 3, 4 and 6 would not form. In general, twins form as thin lamellae and the parent crystal is rarely sheared entirely to the twin orientation. Frank and Thompson[23] pointed out that under some circumstances directions between K_2 and the normal to K_1 may be contracted rather than elongated during twinning: this point is often ignored, even in current literature. The effect is illustrated in Fig. 10(*b*) which shows a small twin of thickness *t* within a large crystal of length x_0. If α is the angle subtended by η_1 and the axis of the crystal, the strain can be calculated and is given by

$$\varepsilon = \left(\frac{t^2 s^2 - 2 t x_0 s \cos \alpha}{x_0{}^2} + 1 \right)^{\frac{1}{2}} - 1 \qquad (10)$$

For $x_0 \sim t$ this expression reduces to a similar form as those given previously, but for $x_0 \gg t$ the expression will be negative for values of $\alpha < 90°$ if $2x_0 \cos \alpha > st$. Thus, for these twins, a contraction may occur in a specimen in which the solution for very thick twins would indicate that an elongation would occur. For twins such as $\{10\bar{1}2\}$ twins in zinc where the shear strain is small the effect is not serious and the zone of 'fortuitous' contractions is only about 4° wide. Alternatively, for $\{112\}$ twins in b.c.c. metals, the zone is about 19° wide.

7. ATOM MOVEMENTS IN TWINNING

This subject has become of increasing interest in recent years because of the realisation that in many of the more complex crystal structures only a few of the atoms are sheared directly to their correct twin positions. In this Section the problem is treated systematically, starting with a simple analysis of the models involved. This is followed by specific examples for particular structure types. In subsequent sections the analysis is extended to an examination of the twin–parent interface and the concept of twinning dislocations is introduced. Finally, the interaction between slip and twinning dislocations is considered.

7.1. Single lattice structures

A single lattice structure is one in which each atom in the crystal is associated with a corresponding site in the Bravais space lattice. Suppose that for a type I twin a primitive vector of the Bravais space lattice in the direction η_2 intercepts a single plane of atoms parallel to the composition plane K_1. It can be shown that all lattice points, and therefore all atoms in the structure, shear directly to the twin position. This is illustrated in Fig. 11 which represents a projection of a single lattice structure in the plane of shear s. Using the notation of Bilby and Crocker[12] this corresponds to $q = 1$ where q is defined as the number of planes parallel to K_1 intersected by a primitive vector parallel to η_2. An alternative notation used by Bevis and Crocker[10] refers to the parameter m which is the reciprocal of the fraction of atoms which move to their final positions in the Bravais space lattice as a result of the twinning shear. For the example, in Fig. 11, therefore, $m = 1$. For the reciprocal mode, Fig. 11(b), which is a type II twin, $q = 1$ corresponds to the number of K_2 planes intersected by the primitive vector along the direction η_1 and all atoms are sheared directly to their correct twin position, i.e. $m = 1$.

For most twins the primitive vectors along the direction η_2 (or η_1) intersect more than one plane parallel to K_1 (or K_2) so that during twinning not all of the lattice points are displaced directly to their positions in the twin. With such twins, the Bravais lattice can be restored only by further inhomogeneous displacements known as *shuffles*. The need for atomic shuffles during twinning has been known for some time, but a meaningful mechanistic analysis has only been developed in recent years, see Kiho,[24,25]

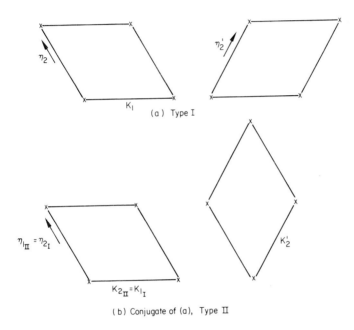

Fig. 11. Plane of shear diagram for a single lattice structure, with $m = 1$, illustrating the displacement of lattice points by the twinning shear: (*a*) type I twin, (*b*) type II twin.

Jaswon and Dove,[26-28] Bilby and Crocker.[12] An example of a twinning process in which $m = 2$ is illustrated in Fig. 12 for type I and type II twins, *cf.* Fig. 11. Associated with each parallelogram defined by either K_1 or η_1 there are a total of four atoms. Of these, only the corner and centre atoms shear directly to their correct twin positions, the remaining two atoms have to shuffle. In this example, the direction of the shuffles is parallel to K_1 and their movement is such that the vector sum of all the shuffles is zero. It should be noted that when shuffling does occur to

restore the structure the magnitudes of the individual shuffles in a particular mode are not necessarily identical to those in the reciprocal mode
even though the twinning shears must be equal.

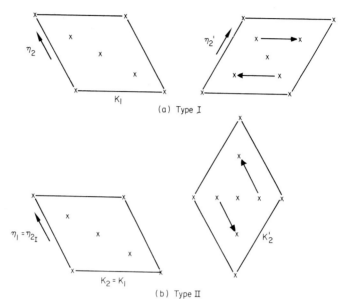

(a) Type I

(b) Type II

Fig. 12. Plane of shear diagram for a single lattice structure, with $m = 2$
illustrating the displacement of the lattice points by the twinning shear and the
shuffles required to restore the original structure: (a) type I twin, (b) type II twin.

7.2. Double lattice structures

In a double lattice structure each site of the Bravais space lattice has
two atoms associated with it. The crystal structure consists of two inter-
penetrating single lattice structures of the same type. The structure is said
to be centro-symmetric. The orientation relationships for type I and type II
twins, discussed earlier, apply equally to twins in double lattice structures.
With these structures many crystal planes assume a corrugated nature
and for convenience each pair of atoms at the Bravais lattice site are
considered to form a *motif pair*, which shears initially as a rigid unit
(Jaswon and Dove[27]). This is illustrated in Fig. 13 for a twin with $m = 1$.
The plane of shear passes through the centre of symmetry of each motif
pair; the atoms in each motif unit lie on either side of the plane of shear so
that the atoms marked $\oplus$ lie above this plane and the atoms marked $\ominus$ lie
below the plane. After shear, Fig. 13(b), all the Bravais lattice points lie

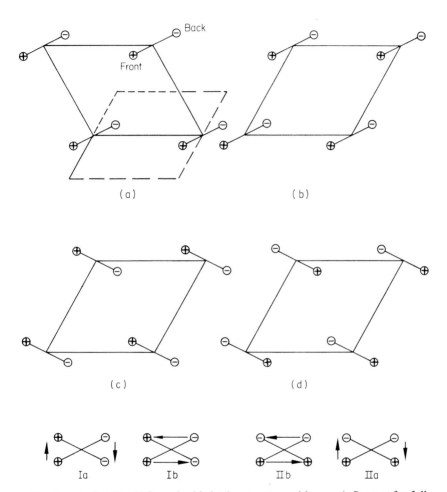

Fig. 13. As for Fig. 11 for a double lattice structure with $m = 1$. See text for full description.

in their correct orientation but none of the atoms are in positions such that the crystal structure is restored. This will only be accomplished by shuffles of the motif units which, for this particular example, remain unbroken. For type I twins the final atomic arrangement is shown in Fig. 13(c). The shuffle mechanism must correspond to either Ia or Ib. In mechanism Ia the atoms on either side of the plane of shear move through equal but opposite amounts along the directions normal to K_1. If the motif units lie close to K_1 this type of shuffle is most probable (Jaswon and Dove[27]). For the shuffle mechanism Ib, also proposed by Jaswon and Dove, the atoms of the motif pair move through equal and opposite amounts in directions parallel to K_1. This mechanism is favoured when the motif unit lies close to the direction normal to K_1. If the twin in Fig. 13 is type II the atoms must shuffle to the final positions by either mechanism IIa or IIb. Mechanism IIa will be relatively simple if the motif pair lies close to η_1, and mechanism IIb will be preferred if the motif pair lies close to a direction perpendicular to η_1.

Again, it is important to note that, although the magnitude of s is identical for a particular twin and its conjugate, the absolute magnitude of the shuffles are, in general, not equivalent. For a particular type I twin the atom shuffles result in the motif unit being reflected through the composition plane K_1. For the conjugate mode, which is a type II twin, the shuffles result in a reflection of the motif unit through a plane which is perpendicular to its direction of shear. Thus, the magnitude of the shuffles in the twin and its conjugate will only be identical in special cases.

For twins with $m > 1$ the shuffle mechanisms are more complicated because not all the Bravais lattice sites shear to their correct twin positions. The motif pairs associated with the Bravais lattice sites shuffle in a *disruptive manner*, i.e., the motif units are not retained during the shuffle, although the vector sum of the shuffles within the 'multiple' cell unit always remains zero. This is illustrated in Fig. 14 for a twin in a double lattice structure with $m = 2$ (Crocker[14]). The atom arrangements before shear are shown in Fig. 14(a) and immediately after shear in Fig. 14(b). In Figs. 14(c) and 14(d) the final positions of the atoms for type I and type II twins are shown. For the non-disruptive shuffles the atom movements are similiar to those illustrated in Fig. 13. The most probable atom movements for the disruptive shuffles are shown in Fig. 14, Ic or Id, for type I twins and IIc and IId for type II twins. For type I twins the disruptive shuffle Ic is preferred if the motif unit lies close to η_1 and shuffle mechanism Id is preferred if the motif unit is parallel to K_1 and is perpendicular to η_1. For type II twins the shuffles of IIc are minimised if the motif unit lies in K_1.

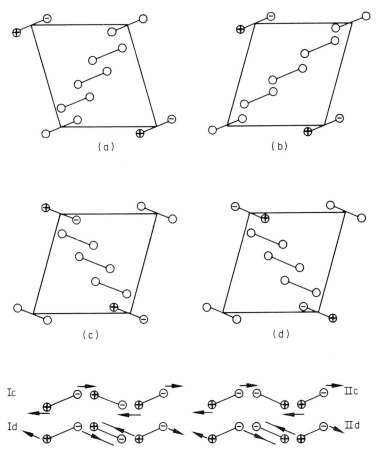

Fig. 14.   As for Fig. 12 for a double lattice structure with $m = 2$. See text for full
description.

The shuffles of IId are always more complex than IIc and, in consequence,
are improbable.

8.   PREDICTION OF THE TWINNING ELEMENTS

Several papers have been written on this subject, see Kiho,[24,25] Jaswon
and Dove,[26-28] and, in particular, Bilby and Crocker[12] and Bevis and
Crocker.[10] The prediction of twinning elements is simplest in single
lattice structures. Some of the general principles involved in the prediction

of twinning modes following the treatment of Jaswon and Dove are
described below. As the treatments by Bilby and Crocker, and Bevis and
Crocker require the extensive use of tensor analysis they will not be discussed here. However, it should be noted that these treatments are, as yet,
the most rigorous and the most general for predicting twinning modes.
In particular, the latter does not assume an orientation relationship between
twin and parent lattices and the former, as well as giving important
relationships between the twinning elements, gives general expressions for
the magnitudes of shuffles associated with type I and type II twinning.
This is illustrated in Fig. 15 for the $m = 1$ case. Suppose that the potential

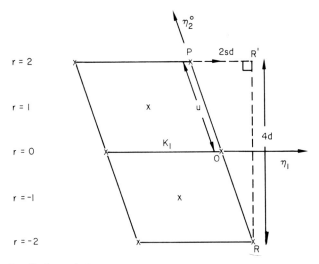

Fig. 15.   Prediction of the twinning elements. Plane of shear diagram for a
single lattice structure with $m = 1$.

interface between the twin and parent is the plane rational K_1 at $r = 0$. On
twinning, the planes parallel to K_1, namely $r = 1$, $r = 2$, etc., shear along
the direction η_1 until they are mirror reflections of the planes $r = -1$,
$r = -2$, etc., through K_1 at $r = 0$. Thus, the primitive lattice vector $\mathbf{u}$
along OP, which is parallel to η_2, shears to OR' in the twin. Since R and R'
are mutual reflections of one another

$$(2u)^2 = (2sd)^2 + (4d)^2$$

giving

$$u^2 = d^2(s^2 + 4) \qquad (11)$$

where d is the spacing of the K_1 planes. Following the reasoning of Jaswon and Dove, $\mathbf{u}$ must be greater than $\mathbf{c}$ the smallest lattice vector of the structure. Thus the expression

$$u^2 \geq c^2 \leq d^2(s^2 + 4)$$

becomes a necessary condition for twinning. Rearranging, this becomes

$$d^2 \geq \frac{c^2}{s^2 + 4} \tag{12}$$

In general, only twins with small shears will be operative so that we can impose the restriction $s \leq 1$, and providing this condition is fulfilled, the expression can be simplified to

$$d^2 \geq \frac{c^2}{5} \tag{13}$$

As an example, consider the *body-centred cubic* lattice. For possible K_1 planes having indices (hkl)

$$d^2 = \frac{I^2 a^2}{h^2 + k^2 + l^2}$$

where I has the values given in Table 1 and a is the lattice parameter of the cubic unit cell. In the body-centred cubic structure the smallest lattice

TABLE 1

VALUES OF I FOR DIFFERENT LATTICE TYPES

Cell	Plane type	I
Primitive	all values of hkl	1
Body centred	$h + k + l$ even	1
	$h + k + l$ odd	$\frac{1}{2}$
Base centred	$h + k$ even	1
	$h + k$ odd	$\frac{1}{2}$
Face centred	h, k and l odd	1
	h, k and l mixed	$\frac{1}{2}$

vector has a magnitude $c^2 = 3a^2/4$. Thus, the possible K_1 planes must satisfy the inequality

$$\frac{I^2}{h^2 + k^2 + l^2} \geq \frac{3}{20}$$

The only planes meeting this requirement are {100}, {110} and {211}. However, {100} and {110} are symmetry planes so that the only possible composition planes for twins in a body-centred cubic structure having $m = 1$ and $s \leq 1$ are of the type $\{211\}$. For this composition plane, the minimum value of s, which is obtained when $u = c$, can be obtained from eqn. (12). Thus,

$$s^2 = \frac{c^2}{d^2} - 4 = \tfrac{1}{2}$$

and

$$s = \sqrt{\tfrac{1}{2}}$$

Since $\mathbf{c} = 1/2\langle 111 \rangle$ in a body-centred cubic lattice the predicted twinning elements for minimum shear are $K_1 = \{211\}$ and $\eta_2 = \langle 111 \rangle$. These are in complete agreement with numerous experimental observations. By similar arguments the twinning elements in other single lattice structures can be predicted. These predictions are presented in Table 2. With the exception of the face-centred rhombohedral structure, all the twinning elements in Table 2 have been confirmed conclusively by experiment. On

TABLE 2

PREDICTED TWIN MODES IN SINGLE LATTICE STRUCTURES

Structure	K_1	η_1	K_2	η_2	S	s	m
B.C.C.	112	$\bar{1}\bar{1}1$	$\bar{1}\bar{1}2$	111	$1\bar{1}0$	$1/\sqrt{2}$	1
F.C.C.	111	$11\bar{2}$	$11\bar{1}$	112	$1\bar{1}0$	$1/\sqrt{2}$	1
F.C. tetragonal	101	$10\bar{1}$	$10\bar{1}$	101	010	$c/a - a/c$	1
F.C. rhombohedral	110	001	001	110	$1\bar{1}0$	$\dfrac{2\sqrt{2}w^{*}}{\sqrt{(1 + 2w)(1 - w)}}$	1

* $w = \cos \alpha$.

the evidence obtained by Andrade and Hutchings[29] using crystalline mercury, Jaswon and Dove[26] concluded that the prediction of an $\{011\}$ twin composition plane for a face-centred rhombohedral structure was correct, and that the conjugate $\{001\}$ mode was not operative since this was the predominant slip plane. Recent work by Crocker et al.,[30] however, disproves both the determination of the habit plane of mechanical twins and the plane of slip in mercury. These workers conclude that the predominant mode of twinning was an irrational type II mode with elements

$$K_1 = `\{\bar{1}35\}', \qquad K_2 = \{\bar{1}11\}, \qquad \eta_1 = \langle \bar{1}21 \rangle, \qquad \eta_2 = `\langle 0\bar{1}1 \rangle',$$

$$s = 0{\cdot}633 \qquad (14)$$

the type I conjugate mode was not detected, $\{\bar{1}11\}$ being the operative slip plane. The reason for the preference of this mode over that predicted by Jaswon and Dove is not clear, particularly since the predicted twin has a smaller shear. However, it is possibly significant that, if the face-centred rhombohedral cell is allowed to degenerate to a face-centred cubic lattice, the observed twinning elements are identical with those predicted for this structure.

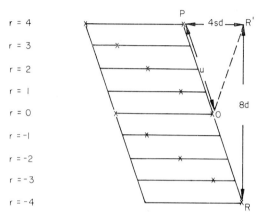

Fig. 16.   As for Fig. 15 with $m = 2$.

The above approach can be used to predict the twinning elements in single lattice structures with $m > 1$. An example is illustrated in Fig. 16 for $m = 2$. Following the method of Fig. 15

$$2u^2 = (4sd)^2 + (8d)^2$$

which reduces to

$$u^2 = 4d^2(s^2 + 4)$$

As before, a necessary condition for the twin to form along a particular (hkl) plane can be expressed as

$$d^2 \geq \frac{c^2}{4(s^2 + 4)}$$

and if $s \geq 1$

$$d^2 \geq \frac{c^2}{20}$$

In applying this inequality to a specific single lattice type it becomes evident that many more composition planes are possible than for $m = 1$.

Of these, many may involve comparatively small shears. For example, in the face-centred cubic structure, twins having the same composition planes as for $m = 1$, but involving only half of the shear, are, in principle, quite possible; the full elements of such a twin would be

$$K_1 = (111), \qquad K_2 = (\bar{1}\bar{1}3), \qquad \eta_1 = [\bar{1}\bar{1}2], \qquad \eta_2 = [332],$$

$$s = \tfrac{1}{2}\sqrt{2} \qquad (15)$$

However, Bilby and Crocker[12] have pointed out that, although the inequality is a necessary condition for twinning, it is not a sufficient condition, particularly when atom shuffles are involved. Thus for twins with $m > 1$ the absolute magnitudes of the shuffles must also be taken into account. Figure 12 shows that the magnitude of the shuffles may be considerable, and, in fact, this factor appears to dominate the choice of twinning elements, since no twins with $m > 1$ have been detected experimentally in single lattice structures. However, in double lattice structures this situation can be reversed because of the smaller shuffles involved. Thus, Jaswon and Dove[27] pointed out that for the diamond structure, which is a double lattice structure of the face-centred cubic type, the above twinning mode having $K_1 = (111)$ and $\eta_2 = [332]$ is likely to be most favoured, since it involves only small shears and small shuffles, cf. Fig. 14. When compared with the $m = 1$ mode for a double face-centred cubic lattice, cf. Fig. 13, such that $K_1 = (111)$ and $\eta_2 = [112]$, it can be seen that the magnitudes of the shuffles are comparable, whereas, the shear is doubled.

In summary, it is clear that in predicting the twin modes, the magnitudes of the shuffles and the twinning shear must be taken into account and, in most cases, the operative twin modes in double lattice structures will not be the same as those in single lattice structures with the same Bravais space lattice. One apparent exception to this is twins which form in the face-centred rhombohedral double lattice structure of antimony, bismuth and arsenic which exhibit the same twinning mode as those predicted by Jaswon and Dove[27] for the single lattice structure.

An example of the combined effect of small shears and small shuffles can be found in the explanation of the two conjugate twin modes of white tin. The structure is normally regarded as a double lattice body-centred tetragonal in which the full twinning elements are

$$K_1 = \{101\}, \qquad K_2 = \{\bar{3}01\}, \qquad \eta_1 = \langle \bar{1}01 \rangle, \qquad \eta_2 = \langle 103 \rangle,$$

$$s = 0{\cdot}113, \qquad m = 2$$

By changing the reference axes to that of a face-centred tetragonal structure this mode can be compared with that of the $\{10\bar{1}\}$ twin listed in Table 2. The twinning elements in this new lattice are

$$K_1 = (111), \qquad K_2 = (\bar{3}\bar{3}1), \qquad \eta_1 = \bar{1}\bar{1}2, \qquad \eta_2 = [116],$$
$$s = 0.113, \qquad m = 2 \qquad (16)$$

The reason for this new twin mode, compared with the $\{10\bar{1}\}$ mode of the single lattice structure, appears to be two-fold. First, for the $\{111\}$ and $\{\bar{3}\bar{3}1\}$ modes the motif unit can be chosen to lie in the plane of shear, thus simplifying the shuffle mechanism even though $m = 2$, whereas for the $\{10\bar{1}\}$ mode of Table 2 the motif units do not lie in the plane of shear and the shuffles, though non-disruptive, would be quite complicated. Secondly, for the low axial ratio of the white tin tetragonal structure, the $\{111\}$ mode is found to require a much smaller shear than the $\{10\bar{1}\}$ mode.

The most detailed investigations on the effect of the magnitude of shear and shuffles on the predicted twinning modes have been made by Crocker[14] for the base-centred orthorhombic double lattice structure of α-uranium. Using the inequality expression for possible composition planes, Crocker found that a minimum of 41 possible twins existed, excluding the conjugate modes having a shear $s < 1$ and with $m = 1$ or 2. Of these, 11 modes were compound and a further 12 had three rational elements. The five most important modes were analysed extensively; reference should be made to the original paper for further details. Crocker was able to show a considerable measure of agreement between the experimentally observed twin modes and the predicted modes providing account was taken of shuffles as well as shears.

This approach has also met with reasonable success for the double lattice structure of the close-packed hexagonal metals. Crocker and Bevis (unpublished) found that there were 18 possible twin modes having $s < 1$ and $m = 1$ or 2 for γ in the range 1·5 to 1·9. Seven modes with the smallest shears for $\gamma = 1·6$ are listed in Table 3. Of these the $\{10\bar{1}2\}$ compound twin has the lowest shear for all known values of the axial ratio γ, with the exception of graphite where $\gamma = 2·725$. Moreover, for this twin mode, the shuffle mechanism is particularly simple because the motif unit lies in the plane of shear $\{\bar{1}210\}$. The shuffles are likely to be either type Ia or IIa, Fig. 13, for non-disruptive shuffles, and type Ic or IIc, Fig. 14, for disruptive shuffles. Consequently, these twins are predicted on this mode to form readily and this is confirmed by experiment. The twin of mode 2 in Table 3 has quite a small shear but the shuffles are complex and this mode has not been observed. The twins of modes 3 and 5 are type I,

TABLE 3

SOME PREDICTED HEXAGONAL TWINNING MODES WITH $s < 1$ FOR AXIAL RATIOS IN THE RANGE $1\cdot5 < \gamma < 1\cdot9$. THE IRRATIONAL ELEMENTS ARE DENOTED BY INVERTED COMMAS

	K_1	η_1	K_2	η_2	s	m	Observed
1	$\{10\bar{1}2\}$	$\langle10\bar{1}\bar{1}\rangle$	$\{10\bar{1}2\}$	$\langle10\bar{1}1\rangle$	$(\gamma^2 - 3)/\sqrt{3}\gamma$	2	Cd, Zn, Mg, Co, Zr, Ti, Be
2	$\{\bar{4}221\}$	$\langle2\bar{1}\bar{1}12\rangle$	$\{0001\}$	$\langle\bar{2}110\rangle$	$1/2\gamma$	2	–
3	$\{10\bar{1}1\}$	'$\langle\bar{1}3\bar{2}1\rangle$'	'$\{\bar{1}6\bar{5}3\}$'	$\langle5\bar{1}43\rangle$	$(4\gamma^4 - 17\gamma^2 + 21)^{\frac{1}{2}}/2\sqrt{3}\gamma$	2	Mg
4	$\{20\bar{2}1\}$	$\langle\bar{1}014\rangle$	$\{0001\}$	$\langle10\bar{1}0\rangle$	$\sqrt{3}/2\gamma$	2	–
5	$\{01\bar{1}3\}$	'$\langle\bar{7}52\bar{1}\rangle$'	'$\{\bar{6}511\}$'	$\langle11\bar{2}3\rangle$	$(4\gamma^4 - 17\gamma^2 + 27)^{\frac{1}{2}}/2\sqrt{3}\gamma$	2	Mg
6	$\{\bar{4}310\}$	$\langle25\bar{7}0\rangle$	$\{01\bar{1}0\}$	$\langle\bar{2}110\rangle$	$1/\sqrt{3}$	2	–
7	$\{\bar{2}111\}$	$\langle2\bar{1}\bar{1}6\rangle$	$\{0001\}$	$\langle\bar{2}110\rangle$	$1/\gamma$	1	Zr, Ti, Re, graphite

and twins on such composition planes have been detected experimentally, although a complete verification of the elements has not been made. Reed-Hill[31] suggested, on the basis of experimental observations, that the $\{10\bar{1}1\}$ and $\{10\bar{1}3\}$ twin modes were conjugate to one another. The full twinning elements would then be

$$K_1 = \{10\bar{1}1\}, \qquad K_2 = \{10\bar{1}3\}, \qquad \eta_1 = \langle10\bar{1}2\rangle, \qquad \eta_2 = \langle30\bar{3}2\rangle,$$

$$s = \frac{4\gamma^2 - 9}{4\sqrt{3}\,\gamma}, \qquad m = 4 \qquad (17)$$

The shear for these twins is considerably less than for the twins suggested by Crocker and Bevis ($s \sim 0\cdot6$) but since $m = 4$ the shuffles will be more complex. However, for this mode a motif unit can be chosen to lie in the plane of shear so that the shuffle mechanism will be less complex than a value $m = 4$ would suggest.

The twins of modes 4 and 6 in Table 3 have large shears and have not been observed, even though for mode 4 the motif unit may lie in the plane of shear, so simplifying the shuffles. Finally, mode 7 is of interest since it is the predominant twinning mode in graphite (Friese and Kelly[32]) for which the twinning shear is small, $s \sim 0\cdot367$, because of the high value of γ. Twins having shear magnitudes comparable to those predicted for mode 7 have also been observed in titanium (Rosi, Duke and Alexander[33]), zirconium (Reed-Hill, Slippy and Buteau[34]) and rhenium (Churchman,[35] Jeffery and Smith[36]). Thornton[37,38] has also detected twins of this type in the hexagonal ζ phase in Cu–Ge, Ag–Al, Ag–Sn, and Ag–Sb alloys. Rapperport,[39] however, reported $\{11\bar{2}1\}$ twins in zirconium with a much smaller shear than that predicted for mode 7. He suggested that the twinning elements were $\{11\bar{2}1\}$, $\{1, 1, \bar{2}, 16\}$, $\langle 11\bar{2}6 \rangle$, $\langle 8, 8, \bar{16}, 3 \rangle$ with a shear $s = 4(\gamma^2 - 4)/17$. This mode is most improbable since complex shuffles are involved and only one atom in 17 is sheared to its correct site. Crocker[40] suggested a double twinning mechanism for these latter twins in which a twin with elements $\{11\bar{2}4\}$, $\{11\bar{2}2\}$, $\langle 22\bar{4}3 \rangle$, $\langle 11\bar{2}3 \rangle$ immediately retwins on a second system $\{11\bar{2}4\}$, $\{11\bar{2}2\}$, $\langle 11\bar{2}3 \rangle$, $\langle 22\bar{4}3 \rangle$ to produce a twin having a simple equivalent mode of $\{11\bar{2}1\}$, $\{11\bar{2}8\}$, $\langle 11\bar{2}6 \rangle$, $\langle 44\bar{8}3 \rangle$ which has a small shear $s = 4(\gamma^2 - 2)/9\gamma$. Whether or not such double twinning occurs is a matter for conjecture since no other $\{11\bar{2}1\}$ twins with a small shear have been reported. Despite the large shear of mode 7 this may well be preferred particularly since no shuffles are involved, $m = 1$.

No other mode having $s < 1$ and $m = 1$ or 2 have been observed in close-packed hexagonal structures. However, Rosi, Duke and Alexander[33] reported $\{11\bar{2}2\}$ twins in titanium and Rapperport[41] reported similar twins in zirconium. At low temperatures Rosi, Perkins and Seigle[42] also observed twins in titanium with $\{11\bar{2}4\}$ composition planes. These two twins are cozonal and Kiho[25] suggested that they were conjugate having full elements

$$K_1 = \{11\bar{2}2\}, \qquad K_2 = \{11\bar{2}4\}, \qquad \eta_1 = \langle 11\bar{2}3 \rangle, \qquad \eta_2 = \langle 22\bar{4}3 \rangle,$$

$$s = \frac{2(\gamma^4 - 4\gamma^2 + 4)^{\frac{1}{2}}}{3\gamma} \qquad (18)$$

Rapperport[39] confirmed these elements for $\{11\bar{2}2\}$ twins in zirconium. For both titanium and zirconium the twinning shear is small but $m = 3$ so that many shuffles are required. However, these shuffles are relatively simple since the motif units lie in the composition plane.

9. ATOMIC ARRANGEMENTS AT TWIN INTERFACES

9.1. Coherent interfaces

When the physical interface between the twin and parent lattices is planar and parallel to the composition plane it is called a *coherent* interface. The simplest twin-matrix interfaces are those in single lattice structures in which twinning occurs without shuffles, $m = 1$. The $\{111\}\langle 11\bar{2} \rangle$ twins in face-centred cubic crystals provide an excellent example. The stacking sequence of $\{111\}$ planes in untwinned crystals is . . . ABCABC . . .. The stacking sequences of these planes across a $\{111\}$ twin interface is

... ABCABĊBACBA .... The position of the boundary is shown by Ċ
and this is illustrated in Fig. 17. The diagram also shows that a one layer
twin can be formed from a single stacking fault and can thicken by the
'addition' of single stacking faults on successive {111} planes. Thus, a

one layer twin, Fig. 17(*b*) has a stacking sequence ... ABCAB/Ċ/BCABC ...
and a twin several layers thick has a stacking sequence ... ABCABC/
BĀCBAĊ/ABCABC ..... The one layer twin is an *intrinsic stacking fault*,
having a large energy due to two violations of the next nearest neighbours

in the stacking sequence ...BCBC .... This transforms the face-centred

cubic lattice locally into a close-packed hexagonal structure which has a
characteristic stacking sequence of close-packed planes of ... ABABAB ....
The thicker twin, on the other hand, does not contain a thin layer of
close-packed hexagonal structure and there is only one violation of the

next nearest neighbour configuration at each boundary ... ABC/BA ...

AC/ABC ..., Fig. 17(*a*). Consequently, the fault energy is lower and there

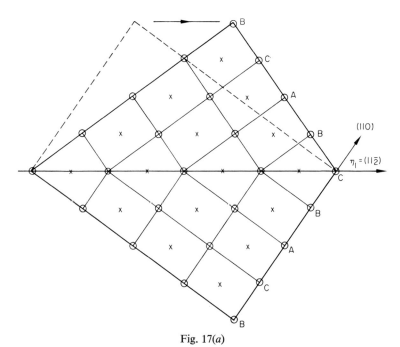

Fig. 17(*a*)

is a sharp reduction in the total twin/parent interface energy as the twin grows from a single layer to a multiple layer twin. This reduction in energy probably plays a significant role in the apparent reduction in stress required to grow a twin compared with the stress for nucleation.

It is interesting to note that in the face-centred cubic structure the $m = 2$ twin mode (eqn. (15)) has exactly the same stacking sequence as the $m = 1$ mode, and the composition planes are identical. For the $m = 2$ mode the twin can be formed by successive additions of extrinsic faults of the type . . . ABCA/C/BCA . . ... Thus, in a region of a face-centred cubic structure having a {111} stacking sequence . . . ABCAB̌CABC . . . a twin can be built up from B̌ by the insertion of extrinsic faults between each successive plane to produce the stacking . . .ABCAB/ACBACBA/ CABC . . .. The $m = 2$ mode has exactly one half the shear of the $m = 1$ mode and the direction of shear is of the opposite sense.

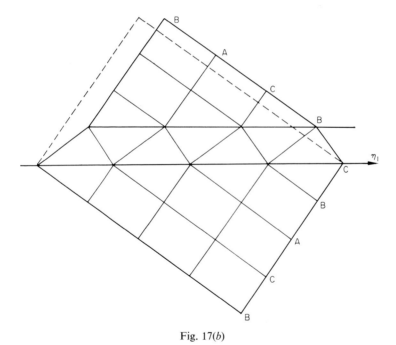

Fig. 17(b)

Fig. 17. (a) Plane of shear diagram for {111} ⟨11$\bar{2}$⟩ twins in face-centred cubic metals illustrating the stacking sequence of {111} planes, before and after twinning, across a single twin boundary. (b). As for (a) showing a one layer twin.

A similar approach can be used for twins in other single lattice structures. Thus, in the body-centred cubic structure, the {112} planes are stacked in a ... DEFABCDEFABC ... sequence and across a twin boundary at F this becomes ... DEFABCDEFEDCBAFE ....

Although twins have not been observed in single lattice structures with $m > 1$ it is useful to consider the stacking arrangements in such structures because twins of this type are observed in double lattice structures. In Fig. 18(a) a twin with $m = 2$ in a single lattice structure is illustrated. This

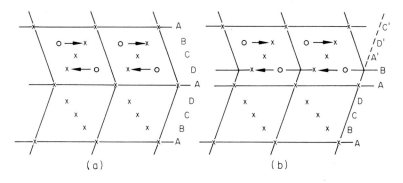

Fig. 18.   Hypothetical plane of shear diagram for a single lattice structure with $m = 2$ showing the shuffles associated with the twinning shear. The stacking sequence is defined in relation to direction η_2. (a) symmetrical twin boundary interface, (b) arrangement of atoms when interface is displaced to an adjacent plane.

could, for example, correspond to a {10$\bar{1}$2} twin in a hypothetical single lattice hexagonal structure. In this example the twin/matrix interface does not contain any atoms which require shuffling. However, if the boundary is displaced to the adjacent parallel composition plane the twin and matrix atoms would no longer be in a symmetrical position with respect to each other across the interface when the crystal structure is restored by shuffles, Fig. 18(b). It follows, that for twins with $m > 1$, the choice of the interface is restricted to those planes containing no shuffled atoms. This point can be demonstrated in another way be considering the stacking arrangement across the twin interface. For the interface in Fig. 18(a) the sequence is ... ABCDÁDCBA ..., providing the sequence is defined along direction η_2, and the sequence in Fig. 18(b) is ... ABCDABA'D'C'B' .... The planes A'D' ... etc. are not in the correct position for twin symmetry.

It follows that, in this example, the minimum twin thickness corresponds
to ... ABCDÁDĆDABCD ... and is two atom layers thick.

In double lattice structures atom shuffles, associated with the motif
pairs, have to occur and in the twin/matrix interface only incomplete
shuffling is possible. Typical interfaces for type I and type II twins, with
$m = 1$, are illustrated in Fig. 19. For the type I twin in Fig. 19(a) the

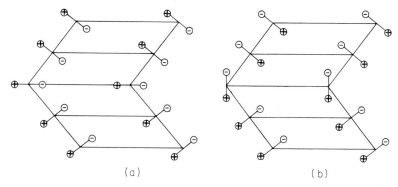

(a)   (b)

Fig. 19.   Plane of shear diagrams illustrating the atomic arrangements at twin
matrix interfaces in double lattice structures.

motif pair lies close to K_1 and the shuffles adopt mode Ia in Fig. 13. Thus,
the number of atoms in the interface is the same as in similar parallel
lattice planes. In Fig. 19(b) the twin is type II and the motif pairs lie close
to a direction perpendicular to η_1. The shuffle mechanism is that of IIb in
Fig. 13 and in the interface the atoms are again displaced from their
equilibrium positions. It follows, that in both examples, the energy of the
twin interface will probably be higher than that of comparable twins in
single lattice structures. Apart from the magnitude of the shuffles it is
clear from this analysis that the shuffle mechanisms operative in twin
growth will influence the stability of the twin/matrix interfaces and are an
additional parameter in predicting the operative twin modes. For twins
with $m > 1$ the arguments can be further extended. Thus, a twin boundary
in such structures would be expected to contain only non-disruptive
shuffles. Interfaces containing disruptive shuffles would be of the non-
symmetrical type and involve large energies.

9.2.   Semi-coherent and non-coherent interfaces
So far only twins which have perfect matching with the parent lattice
across the interface have been considered. Frequently, however, twin

boundaries are curved and are not parallel to K_1. In these circumstances the boundaries must contain some degree of *incoherence*. Such points of incoherence can be described by misfit dislocations usually referred to as *twinning dislocations*. A schematic representation of a twinning dislocation for a single lattice structure containing a twin, $m = 1$, is shown in Fig. 20. In Fig. 20(*a*) a closed circuit around the twinning dislocation is traced

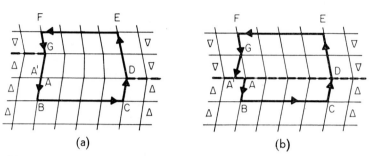

Fig. 20.   Burgers vector of a twinning dislocation: (*a*) section across coherent twin boundary, (*b*) section across a boundary containing a twinning dislocation (after Friedel).[43]

along three invariant directions, η_2' in the twin, η_2 in the matrix and *BC*, which is a rational direction contained in K_1. The same circuit is illustrated in Fig. 20(*b*) across a perfect twin boundary and there is a closure failure *AA'*. The displacement is parallel to η_1 lying in K_1 and defines the vector $\mathbf{b}_t$ of the twinning dislocation. By analogy with other dislocations, twinning dislocations may be curved and adopt edge, screw or mixed characteristics.

The vector $\mathbf{b}_t$ defined in this way for a twin having a rational K_1 is always parallel to η_1. Moreover, since the vector was defined by tracing out a Burgers circuit along rational directions, this vector is constant and independent of the position of the twinning dislocation. As a consequence the twinning dislocation can, in principle, glide along the composition plane and alter the volume of twinned crystal. A lenticular-shaped twin bounded by a semi-coherent interface containing such twinning dislocations is illustrated in Fig. 21. If the dislocations in the boundaries *LM* and *MN* glide to the left the volume of twin increases. The corresponding shear produced defines the magnitude of the twinning dislocation and is given by

$$|\mathbf{b}_t| = sd$$

where s is the twinning shear and d is the interplanar spacing of planes parallel to K_1.

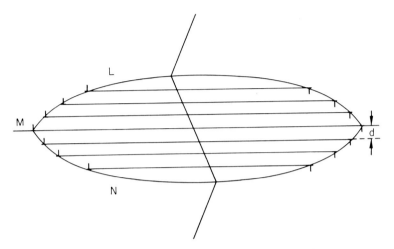

Fig. 21.   Lenticular twin with semi-coherent interface.

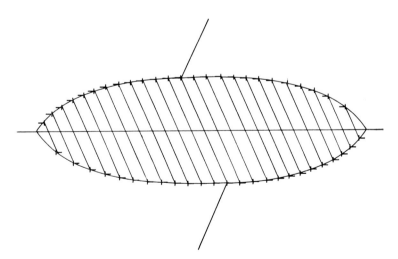

Fig. 22.   Lenticular twin with incoherent interface (after Friedel).[43]

Friedel[43] has pointed out that the twinned volume in Fig. 21 could be produced by a second set of dislocations with Burgers vectors arranged as a wall of parallel edge type dislocations as shown in Fig. 22. This twin will grow freely by glide of the dislocations normal to the boundary plane which is incoherent and, in general, will be irrational. For this reason this mode of a twin interface is particularly useful in considering some of the properties of type II twins which have an irrational composition plane.

The twinning dislocations of body-centred cubic and face-centred cubic

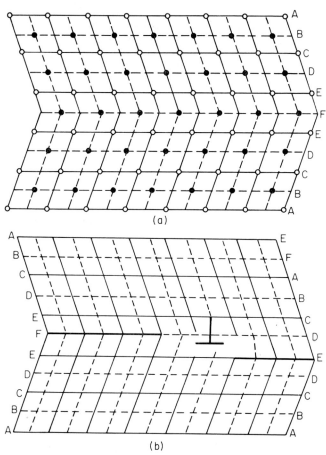

Fig. 23.   Stacking sequence of {112} planes for {112}⟨$\bar{1}\bar{1}1$⟩ twins in a body-centred structure: (a) coherent interface, (b) containing a $\frac{1}{6}$⟨$\bar{1}\bar{1}1$⟩ twinning dislocation.

single lattice structures provide the simplest examples of twin interfaces of the type shown in Fig. 21. As mentioned earlier, the twinning shear in a body-centred cubic lattice is along $\langle\bar{1}\bar{1}1\rangle$ parallel to a $\{112\}$ plane. From the stacking sequence of $\{112\}$ planes across a twin interface, see Fig. 23(a), it can readily be shown that the vector of the twinning dislocation is $\frac{1}{6}\langle\bar{1}\bar{1}1\rangle$. This dislocation is illustrated in Fig. 23(b). It is worth noting that a displacement of $\frac{1}{6}\langle11\bar{1}\rangle$ would not produce a twin but a displacement of $\frac{1}{3}\langle11\bar{1}\rangle$, which corresponds to a shear in the opposite sense to $\frac{1}{6}\langle\bar{1}\bar{1}1\rangle$ of twice the magnitude produces the same twin. However, the latter mode has not been observed experimentally, presumably because of the excessively large shear involved.

In the face-centred cubic lattice the twinning dislocations for the $m = 1$ twin mode correspond to Shockley partial dislocations $\frac{1}{6}\langle11\bar{2}\rangle$. For the $m = 2$ twin mode (eqn. (15)) the twinning dislocations are composed of two Shockley partial dislocations lying on adjacent planes so that their vectors are given by

$$\tfrac{1}{6}[\bar{2}11] + \tfrac{1}{6}[1\bar{2}1] \rightarrow \tfrac{1}{6}[\bar{1}\bar{1}2]$$

The sequence of events as the twin interface moves is illustrated in Fig. 24. The multiple cell defined by the primitive vectors $\frac{1}{2}[\bar{1}\bar{1}2]$ and $\frac{1}{2}[332]$ is shown in Fig. 24(a). After plane A is displaced by the twinning vector

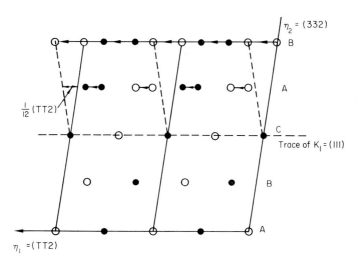

Fig. 24(a)

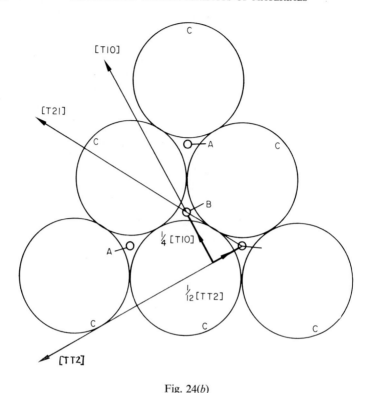

Fig. 24(b)

Fig. 24   (a). Plane of shear diagram for {111}[$\bar{1}\bar{1}2$] twin modes ($m = 2$) in face-centred cubic structures. The macroscopic shear displacement is equivalent to a displacement of $\frac{1}{12}[\bar{1}\bar{1}2]$ on every successive atomic plane parallel to K_1. The open circles lie in the plane of the diagram and the closed circles in adjacent parallel planes. (b). Shuffle mechanism, displacement $\frac{1}{4}[\bar{1}\bar{1}0]$, which accompanies the macroscopic shear. The C atoms lie in the {111} composition plane and the atoms resting in A positions move to B position.

$\frac{1}{12}[\bar{1}\bar{1}2]$ the atoms are in non-symmetrical positions, so the interfacial energy is high. By a shuffle of all the atoms in layer A by the vector $\frac{1}{4}[\bar{1}10]$, i.e. in the plane K_1 normal to the plane of the diagram, the twin symmetry is obtained. The details of these movements are shown in Fig. 24(b) which is a projection onto a $K_1 = (111)$ plane; the C atoms represent the atoms in the composition plane. This shows that the effective displacement of the A atoms $\frac{1}{12}[\bar{1}\bar{1}2] + \frac{1}{4}[\bar{1}10]$ is $\frac{1}{6}[\bar{2}11]$. If the twin now grows by a further (111) layer the shuffle takes place in the opposite direction and the effective displacement $\frac{1}{12}[\bar{1}\bar{1}2] + \frac{1}{4}[1\bar{1}0]$ is $\frac{1}{6}[1\bar{2}1]$. Thus,

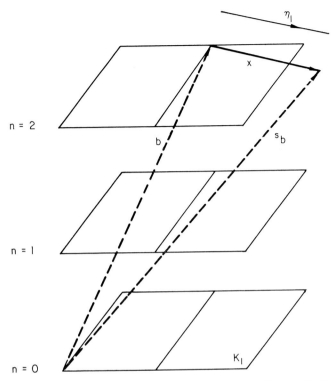

Fig. 25.   Non-conservative transformation associated with a twinning shear.

by the operation of $\frac{1}{6}[\bar{2}11]$ and $\frac{1}{6}[1\bar{2}1]$ Shockley partials, the twinning shear is produced.

With the non-cubic single lattice structures the stacking of the planes parallel to the composition planes, in general, contain no repeat symmetry so that the twinning dislocations do not have a simple vector. Thus, for the face-centred tetragonal structure the net twinning vector operating for the predicted mode in Table 2 on each successive atomic layer is

$$\mathbf{b}_t = \frac{2(\gamma^2 - 1)}{(\gamma^2 + 1)} \langle 10\bar{1} \rangle$$

where γ is the axial ratio of the crystal. For the face-centred rhombohedral structure the vector is

$$\mathbf{b}_t = \frac{1}{4w} \langle 100 \rangle$$

where $w = \cos \alpha$, and α is the angle subtended by any two principal directions of the primitive lattice.

In double lattice structures, the twinning dislocations can be described in a similar way, except that the composition plane is now corrugated, due to the motif units. Because of the asymmetry which twins with $m > 1$ produce on certain interfacial planes, as already described, it follows that the twinning dislocation will prefer to exist in arrays, rather than separately, to minimise the total twin matrix interface energy. Thus for twins with $m = 2$ the twinning dislocations will prefer to exist in pairs, and for twins with $m = 3$, in groups of three. Westlake[44] has suggested that the groups of twinning dislocations can be regarded as a single zonal dislocation of multiple vector. For twins with large m such arrangements would probably have an impossibly large elastic strain energy and the dislocations would probably separate. The extent of separation would depend on the mutually repulsive forces and the energy of the unsymmetrical interfaces which would be formed by separation. As before, the vector of the twinning dislocations in double lattice structures is not a simple fraction of the lattice vector, but can be obtained from the expression $|\mathbf{b}_t| = sd$. Thus, as an example, the vector of the $\{10\bar{1}2\}\langle\bar{1}011\rangle$ twinning dislocations of the hexagonal lattice is

$$\mathbf{b}_t = \frac{\gamma^2 - 3}{2(\gamma^2 + 3)} \langle\bar{1}011\rangle$$

where γ is the axial ratio ($\gamma > \sqrt{3}$).

10.   THE INCORPORATION OF SLIP DISLOCATIONS BY TWINS

The basic material covered in this chapter can be used to analyse the interrelation, in geometrical terms, between slip dislocations and twins. The problem is too complex to be described in general terms and this Section will be restricted to some simple examples. Two main types of interaction arise; firstly the intersection of a twin by a gliding dislocation, and secondly, the intersection of a pre-existing dislocation by a growing twin. It will be apparent from the analysis that twins are not necessarily obstacles to the propagation of slip dislocations and that slip twin intersections may provide a good mechanism for twin broadening.

The passage of a slip dislocation causes two halves of a crystal to move past each other over a distance equal to the Burgers vector $\mathbf{b}$ of the dislocation, so that, when a dislocation crosses a twin boundary a step is formed in the boundary. The height of the step can be described by the number of K_1 planes intersected by **b**. If the indices of K_1 are (HKL) and the components of **b** are $[pqr]$ the step height may be shown to be equal to $(pH + qK + rL)d_{HKL}$, where d_{HKL} is the interplanar spacing of K_1 planes and $(pH + qK + rL) = n$ is integral when **b** is a lattice vector and K_1 is a rational plane. The components of **b** referred to the twin basis, $[pqr]_T$ can be obtained from $[pqr]_M$, the components of **b** referred to the parent basis, by using the transformation formulae, eqn. (3), where HKL are the indices of K_1 and UVW are the indices of the direction normal to K_1, hereafter called eqns. (3A). This is called a *conservative* transformation, because the magnitude and direction of $(\mathbf{b})_T$ and $(\mathbf{b})_M$ are identical. However, unless the Burgers vector of the slip dislocation is parallel to the twin plane K_1, the indices $[pqr]_T$ and $[pqr]_M$ will be different.

As an example, the $(\bar{1}12)$ $(1\bar{1}2)$ $[1\bar{1}1]$ $[\bar{1}11]$ twinning mode in body-centred cubic crystals will be considered. The $\pm\frac{1}{2}(\bar{1}\bar{1}1)_M$ screw dislocation gliding into the coherent $(1\bar{1}2)$ twin boundary transforms by eqns. (3A), for the orientation relationship of reflection in $(\bar{1}12)$, to $\mp\frac{1}{6}(151)$. The magnitude of these vectors are equal. For dislocations undergoing the transformation the resultant dislocations in the twin will, in general, be imperfect, that is $(\mathbf{b})_T$ will not be an integral multiple of the primitive lattice vector. However, the propagating dislocation may decompose into a definite number of twinning dislocations and, in the example, where $m = 1$ to a *perfect* dislocation, which may propagate into the twin.

The number of twinning dislocations is determined by the number of K_1 planes intersected by **b**. These features are illustrated in Fig. 25. In this case **b** intersects two K_1 planes and $^s\mathbf{b}$ represents **b** after shear by the twinning shear, and **x** is a vector which is parallel to η_1, $\mathbf{b} + \mathbf{x} = {}^s\mathbf{b}$ so that $(\mathbf{b})_M = (\mathbf{b})_T = ({}^s\mathbf{b})_T - \mathbf{x}$. The components or indices of $({}^s\mathbf{b})_T$ may be obtained from $(\mathbf{b})_M$ using the *non-conservative* transformation represented by eqns. (3) where HKL are K_1 and UVW are η_2 since these equations define the components of a direction in the parent lattice which has been sheared, the indices of the resultant being referred to the twin basis. The vector **x** has a magnitude $n \cdot s \cdot d_{HKL}$ and is parallel to η_1 so that $\mathbf{x} = n\mathbf{b}_t$ where $\mathbf{b}_t$ is the twinning dislocation, and hence $(\mathbf{b})_M = ({}^s\mathbf{b})_T - n\mathbf{b}_t$. From the form of the transformation eqns. (3) it is clear that for an $m = 1$ mode and a perfect slip dislocation gliding into a twin boundary, this dislocation can always decompose into twinning dislocations and a perfect dislocation. This resultant dislocation may, or may not, be a slip dislocation; in the latter case the dislocation may be able to decompose into a slip dislocation which propagates in the twin under suitable conditions.

The results obtained above are quite general for $m = 1$ compound twinning modes in all crystal systems, but some qualification is required when type 1 and type II twinning modes are considered and, in particular, for $m > 1$ modes, which necessitate the introduction of zonal dislocations.

The resultant product dislocation of the incorporation of a slip dislocation by a *growing twin* is identical to that described above. In this case, the slip dislocation interacts with propagating twinning dislocations, again resulting in the vector diagram and the equation $(\mathbf{b})_M = (\mathbf{b})_T - n\mathbf{b}_t$. The resultant Burgers vector of the dislocation formed in the twin, is the same in both cases.

Sleeswyk and Verbraak[45] and Ishii and Kiho[46] have considered the incorporation of slip dislocations with deformation twins in body-centred cubic metals and β-tin respectively. Some of the interactions considered by Sleeswyk and Verbraak are given below. Following the procedures described above the interaction of the $\frac{1}{2}[1\bar{1}1]_M$ and $\frac{1}{2}[\bar{1}11]_M$ slip dislocations with the $(\bar{1}12)$ body-centred cubic $m = 1$ twin are considered. For these cases $n = 0$ and 2 respectively, so that

(a) $\pm\frac{1}{2}[1\bar{1}1]_M = \pm\frac{1}{2}[1\bar{1}1]_T$

(b) $\pm\frac{1}{2}[\bar{1}11]_M = \pm\frac{1}{6}[11\bar{5}]_T = \pm\frac{1}{2}[1\bar{1}\bar{1}]_T \mp 2 \times \frac{1}{6}[1\bar{1}1]$

For case (a) the Burgers vector of the slip dislocation lies in the twin plane, there is no step in the boundary, so the dislocation may pass through the twin boundary unaltered. The decomposition of the $\frac{1}{2}[\bar{1}11]_M$ dislocation on entering the twin is illustrated in Fig. 26(a) and (b). Following Sleeswyk and Verbraak, the three-pronged symbol represents $\frac{1}{2}[\bar{1}11]_M$ dislocation. As indicated in reaction (b), two twinning dislocations are left in the boundary, thus providing a method of growth in the transverse direction of the twin. As the propagating $\frac{1}{2}[1\bar{1}\bar{1}]_T$ dislocation is incorporated in the parent, the inverse reaction will apply, as indicated in Fig. 26(c), because no decomposition other than $(\mathbf{b})_M \rightarrow (\mathbf{b})_T - 2n\mathbf{b}_t$ has occurred. This arises because all the transformation formulae (3) and (4) of Section 3 are identical for the twin-parent and parent-twin transformation of indices, and is a consequence of the orientation relationships associated with type I and type II twinning modes.

11. CONCLUSION

As indicated in the introduction, no attempt has been made to give a comprehensive treatment of mechanical twinning. This would require considerably more space than is available. The preceding Sections have

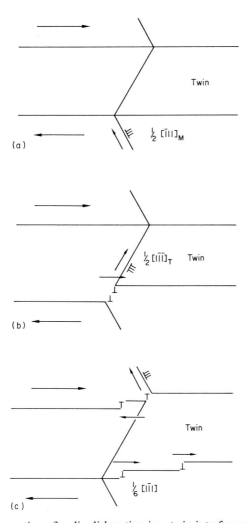

Fig. 26. Incorporation of a slip dislocation in a twin interface producing a new slip dislocation in the twin and two twinning dislocations at each twin-matrix interface (after Sleeswyk and Verbraak). [45]

concentrated on the structural characteristics of mechanical twins as they exist rather than on a description of the phenomena associated with mechanical twinning, which would include:

(1) A description of the conditions of stress, strain rate, and the temperature under which they form.

(2) The material parameters such as purity, orientation, grain size, dislocation density and distributions, which affect whether or not, under a given set of conditions, twins will form.

(3) The kinetics of twin growth.

(4) The specific mechanisms of twin nucleation and growth, in a variety of crystal structures.

(5) The contribution of the plastic strain, which accompanies twins, to the plastic deformation of solids, whether it be in the conventional conditions of a simple uniaxial tensile, or in the highly complex conditions associated with the secondary shape deformations in a martensitic transformation.

(6) The mechanisms of relaxation of the large local strains which accompany twins, particularly when they are held up at barriers, and the role of these relaxation processes on the kinetics of twin growth, the morphology of twins and nucleation cracks.

It could be argued that many generalisations, about the topics listed above, can be deduced, or inferred, from the precise and formal description, which has been adopted in the remainder of the chapter. It is certainly true that any model for, or description of, the phenomena mentioned must be in agreement with the formal description, and thus, this chapter, may be regarded as a starting point for an understanding of these phenomena. There is a vast literature on the topics mentioned, much of which provides a valuable phenomenological background to mechanical twinning. However, an adequate description would require two or three additional chapters. A qualitative survey of these phenomena would be of little value and the reader is, therefore, referred to the technical and scientific literature.

12. ACKNOWLEDGEMENT

We would like to express our grateful appreciation of the contribution made to this chapter by Dr M. J. Bevis, who has not only read and corrected the original text, but has also made a number of valuable additions.

The manuscript for this chapter was submitted to the editor in April 1968.

13. REFERENCES

1. Friedel, G. (1926). *Leçons de Cristallographie*, Berger-Levzault, Paris.
2. Cahn, R. W. (1954). *Adv. in Phys.*, **3**, 363.
3. Hall, E. O. (1954). *Twinning and Diffusionless Transformations in Metals*, Butterworth, London.
4. Christian, J. W. (1965). *Theory of Transformations in Metals and Alloys*, Pergamon, Oxford.
5. Klassen-Neklyudova, M. V. (1964). *Mechanical Twinning in Crystals*, Consultants Bureau, New York.
6. Reed-Hill, R. E., Hirth, J. P. and Rogers, H. C. (Eds.) (1964). *Deformation Twinning*, Gordon & Breach, New York.
7. Johnsen, A. (1914). *Neues Jahrb. Mineral Geol.*, **39**, 500.
8. Niggli, P. (1929). *Z. Kristallogr.*, **71**, 413.
9. Crocker, A. G. (1962). *Phil. Mag.*, **7**, 1901.
10. Bevis, M. J. and Crocker, A. G. (1968). *Proc. Roy. Soc.*, **A304**, 123.
11. Schmid, E. and Boas, W. (1935). *Kristallplastizität*, Springer-Verlag, Berlin.
12. Bilby, B. A. and Crocker, A. G. (1965). *Proc. Roy. Soc.*, **A228**, 240.
13. Cahn, R. W. (1953). *Acta Met.*, **1**, 49.
14. Crocker, A. G. (1965a). *J. Nucl. Mats.*, **16**, 306; (1965b). *Trans. Met. Soc.*, *AIME*, **233**, 17.
15. Bevis, M. J. (1968a). *J. Nucl. Mats.*, **26**, 235.
16. Bevis, M. J. and Crocker, A. G. (1969). *Proc. Roy. Soc.*, **A313**, 509.
17. Pabst, A. (1956). *Bull. Geol. Soc. Am.*, **66**, 897.
18. Andrews, K. W. and Johnson, W. (1955). *Brit. J. Appl. Phys.*, **6**, 92.
19. Bevis, M. J., to be published.
20. Liu, Y. C. (1963). *Trans. Met. Soc.*, *AIME*, **227**, 775.
21. Bevis, M. J. and Swindells, N. (1967). *Phys. Status Solidi*, **20**, 197.
22. Greninger, A. B. and Troiano, A. R. (1949). *Trans. AIME*, **185**, 590.
23. Frank, F. C. and Thompson, N. (1955). *Acta Met.*, **3**, 30.
24. Kiho, H. (1954). *J. Phys. Soc. Japan*, **9**, 739.
25. Kiho, H. (1958). *J. Phys. Soc. Japan*, **13**, 269.
26. Jaswon, M. A. and Dove, D. B. (1956). *Acta Cryst.*, **9**, 621.
27. Jaswon, M. A. and Dove, D. B. (1957). *Acta Cryst.*, **10**, 14.
28. Jaswon, M. A. and Dove, D. B. (1960). *Acta Cryst.*, **13**, 232.
29. Andrade, E. N. da C. and Hutchings, P. J. (1935). *Proc. Roy. Soc.*, **A148**, 120.
30. Crocker, A. G., Heckscher, F., Bevis, M. J. and Guyoncourt, D. (1966). *Phil. Mag.*, **13**, 1191.
31. Reed-Hill, R. E. (1960). *Trans. Met. Soc.*, *AIME*, **218**, 554.
32. Friese, E. J. and Kelly, A. (1961). *Proc. Roy. Soc.*, **A264**, 269.
33. Rosi, F. D., Duke, C. A. and Alexander, B. H. (1953). *Trans. Met. Soc.*, *AIME*, **197**, 257.
34. Reed-Hill, R. E., Slippy, W. A. and Buteau, J. L. (1963). *Trans. Met. Soc.*, *AIME*, **227**, 976.
35. Churchman, A. T. (1960). *Trans. Met. Soc.*, *AIME*, **218**, 262.
36. Jeffery, R. A. and Smith, E. (1966). *Phil. Mag.*, **13**, 1163.
37. Thornton, P. H. (1966). *Acta Met.*, **14**, 141.
38. Thornton, P. H. (1966). *Acta Met.*, **14**, 1257.
39. Rapperport, E. J. (1959). *Acta Met.*, **7**, 254.
40. Crocker, A. G. (1963). *Phil. Mag.*, **8**, 1077.
41. Rapperport, E. J. (1955). *Acta Met.*, **3**, 208.
42. Rosi, F. D., Perkins, F. C. and Seigle, S. S. (1956). *Trans. Met. Soc.*, *AIME*, **206**, 115.

43. Friedel, J. (1964). *Dislocations*, Pergamon, Oxford.
44. Westlake, D. G. (1961). *Acta Met.*, **9**, 327.
45. Sleeswyk, A. W. and Verbraak, C. A. (1961). *Acta Met.*, **9**, 917.
46. Ishii, K. and Kiho, H. (1963). *J. Phys. Soc. Japan*, **18**, 1122.

CHAPTER 4

MARTENSITIC STRUCTURES

WALTER S. OWEN AND FREDERICK J. SCHOEN

1. INTRODUCTION

A martensitic transformation produces a change in the shape of a crystal and this is the property by which such transformations are identified. It occurs by the movement of the plane interface between the newly formed martensite crystal and the parent lattice. The movement of the atoms across the interface involves jumps shorter than the lattice diffusion jump distance and consequently martensitic transformations are usually described as being diffusionless. However, in some closely related but more complex transformations one of the atomic species, the solute, diffuses at the same time as the solvent lattice transforms martensitically. The atoms moving across the interface in a pure martensitic transformation have to move in a coordinated manner to ensure that the interface remains plane and Frank[1] has suggested that the term 'military transformation' be applied to the general class of transformations involving such movements.

Martensitic transformations can occur in a wide range of materials, provided the rate of cooling from some annealing temperature to the martensite transformation temperature (M_s) is sufficiently rapid to prevent any diffusional transformation occurring. Many of the pure elements which are polymorphic can be transformed martensitically as well as a large number of alloy systems. Martensitic transformations in some non-metallic systems have been recognised although these have not been studied in detail. The transformation can occur in a great variety of crystal structures and it is not confined to a small number of parent-product pairs of structures. A comprehensive summary of the major features of the martensitic transformations which have been reported to date has been published recently.[2]

The term 'martensitic' was first applied to the transformation which occurs on quenching steels and the product was called martensite. It is now known that this transformation is usually complex, involving some rearrangement of the carbon atoms as well as a martensitic transformation of the iron lattice. Although no longer considered the prototype of a martensitic transformation, the transformation in steel and other alloys of iron is still of interest and the subject of much modern research. The crystallography and the fundamental properties of the transformation have been studied most extensively and systematically in iron–nickel–carbon alloys in the ranges zero to 35 w/o nickel and zero to 1·8 w/o carbon. The examples discussed in this survey will be taken largely from work on these alloys with the addition of some observations on other iron alloys. In researches carried out during the last decade, alloys have been selected from iron–nickel–carbon or a related series such that, under the conditions of temperature and cooling rate imposed, the transformation is truly martensitic and the secondary diffusional effects encountered in quenched steels are avoided. Work on martensitic transformations in alloys of uranium, titanium or copper has demonstrated that all of the concepts which have been developed by the study of iron alloys are widely applicable.

The formal theories of the crystallography, which must be the starting point and the basis of all discussion of the structural aspects of martensitic transformation, grew out of the work of Greninger and Troiano[5] and were developed during the past twenty years largely by three independent groups; Bowles and Mackenzie in Australia, Wechsler, Lieberman and Read in Illinois and Bilby, Christian, Crocker and others in England. The theory can be described and used either by stereographic projections or in terms of matrix algebra. The latter is usually preferred because greater precision can be achieved but stereographic projection will be used extensively throughout this discussion because it often helps considerably when trying to visualise the geometry. Each of the groups has used different mathematical procedures, although all of the procedures are equivalent in the essential features, and each has introduced some special assumptions. Three excellent reviews[3,4,127] of the formal theories, in which the different procedures are set out in detail, and the results are compared with experimental data have been published within the last six years and no attempt will be made here to cover the same ground. Instead, the assumptions and the basic concepts of the general theory will be outlined and, after the briefest description of the manipulation, the major results will be indicated. This will be sufficient to provide a framework for the discussion of degenerate forms of martensite, related structures such as bainite, the nucleation and growth of these structures and, finally, a brief outline of the fundamentals of the plastic deformation behaviour of the products of martensitic transformation.

There are two major classes of martensite morphology; acicular and massive. In acicular-martensite individual martensite plates exist in a matrix of the parent phase. The plates may touch at the tips to form a zig-zag arrangement but otherwise the plates are unconnected (*see* Fig. 3). In massive-martensite the plates form in parallel array to make a block of martensite (*see* Fig. 19). When acicular-martensite forms the transformation seldom goes to completion, but massive-martensite never contains a detectable amount of untransformed parent phase. The terminology, acicular and massive, is not satisfactory and not universally used.* The adjectives describe the appearance of etched sections in an optical microscope. The plates are not needlelike. The blocks are not formed by a massive transformation. However, since no completely satisfactory way of identifying the two types of structure has been proposed the terminology will be continued here.

2.   THE FORMAL CRYSTALLOGRAPHIC THEORY

The definition of a martensitic transformation as one in which there is a change in shape of the crystal is not unambiguous. However, it is the only definition on which a theory of the crystallography of martensitic change can be constructed. Part of a flat polished surface of a parent crystal is tilted when a martensite plate forms. The tilt reveals the location of the plate, and the tilted surface is flat (Fig. 1). This observation caused early investigators to assume that the martensite plate had formed by a simple lattice shear parallel to the habit plane. An important advance was made when, in 1949, Greninger and Troiano[5] showed that application of a single shear to the parent lattice of an iron–nickel–carbon alloy would not produce the lattice of the martensite crystal. They proposed a two-shear mechanism, the first producing the change in shape observed by optical microscopy of the tilts on surfaces polished flat before transformation, and the second occurring by slip within the martensite crystal to produce the observed lattice. Although this reasoning was applied only to the iron–nickel–carbon martensite which Greninger and Troiano studied experimentally, it was soon realised that a single shear mechanism is inadequate to

* Acicular-martensite has been called normal, twinned, high-carbon and lower martensite, while the descriptions blocky, 'lathe', dislocated, low-carbon and upper have been applied to massive-martensite.

Fig. 1.　Surface shears produced by martensite plates showing flat surfaces and the midribs (Hull and Garwood[5 8]).

explain many other martensitic transformations. However, their two-shear mechanism is also unsatisfactory because it does not, in fact, produce the observed change in shape of the lattice. In particular, it does not produce the change in volume which occurs in many martensitic transformations. When a $1{\cdot}0\%$ carbon steel transforms from austenite to martensite there is an increase in volume of about $4{\cdot}3\%$. Nevertheless, the idea that more than one deformation is involved in a martensitic transformation has proved to be important and all theories since Greninger and Troiano have retained the concept of two deformations; one of which produces a change in the shape of the lattice and thus is known as the *lattice deformation* and the other changes the shape of the martensite crystal without altering the shape of the lattice and consequently is known as the *lattice invariant deformation*. The convention of calling the two deformations the first and second shear is sometimes encountered although no time sequence is meant to be implied. This is a survival of the Greninger–Troiano concept but its use is not advisable because the lattice deformation is not a simple shear.

The lattice deformation and the lattice invariant deformation when combined give the *total shape deformation,* the macroscopic shape change

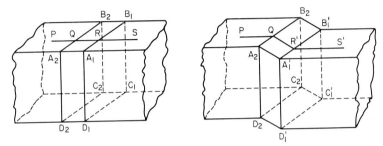

Fig. 2. Representation of a plate of martensite formed from a rectangular block of parent crystal.

which can be measured by measuring the tilt of a surface. Some other important features of the total shape deformation can be deduced by careful microscopic observation of a single plate of martensite embedded in its parent phase. The simplest possible situation, in which the plate of martensite penetrates completely through a single crystal of austenite, is shown in Fig. 2. In real specimens the plate usually ends inside the austenite crystal and some constraint, elastic or plastic, is imposed. However, examination of partially transformed specimens in which the distortions due to the constraints are small shows that a flat surface such as $A_1 A_2 B_2 B_1$ remains plane after the transformation. Secondly, a straight line such as PQRS is turned through two angles but the segment QR' on the surface of the martensite plate is straight. Thus, a plane surface remains plane and a straight line is not distorted by the shape change. From these observations it is concluded that the shape change is homogeneous. Further, lines which are the trace of the intersection of the habit plane with the plane of polish (for example, $A_1'B_1'$ or A_2B_2) are straight, unchanged in length and unchanged in angular orientation relative to the parent crystal. The polished plane $A_1A_2B_2B_1$ is an arbitrary plane containing the trace lines. Thus, the lines $A_1'B_1'$ and A_2B_2 are arbitrary lines in the interface between the parent and the transformed crystal. That is, they are arbitrary lines in the habit plane. Thus, it follows that the habit plane is unrotated and undistorted by the transformation. The habit plane is an *invariant plane*. Of course, the habit plane itself is not an arbitrary plane. As will become clear later, it is a specific but usually irrational plane. The resolution with which the interface traces $A_1'B_1'$ and A_2B_2 and scratches such

Fig. 3. Martensite plates of $\{259\}_P$ habit in an iron–24% nickel–0·5% carbon alloy showing the midrib. Etched transverse section, original ×600 (R. Brook[6]).

as PQRS can be observed is limited and the possibility of some distortion close to the interface cannot be discounted. The Wechsler–Lieberman–Read theory assumes that there is exact matching across the interface but the Bowles–Mackenzie theory introduced a small dilatation which is assumed to be isotropic within the plane of the interface. There is, of course, no reason why more complex distortions are less probable and some of the recent theories have introduced interfacial strains of various types. Figure 2 represents a martensite plate in an unconstrained parent crystal. In a polycrystalline aggregate the martensite plates stop at a grain boundary and assume a lenticular shape to accommodate the elastic constraints imposed by neighbouring grains. The interface is not plane and often the habit plane is distorted both by the accommodation strains and by neighbouring plates. Many martensite plates have a midrib (Fig. 3) which is the trace of the plate at the start of its growth. The scatter in the measurement of the indices of the habit-plane of martensite plates in polycrystals can be reduced but not eliminated by measuring the trace of the midrib.

The total shape deformation, as observed by optical microscopy, is homogeneous. It contains a shear component and also accommodates a volume change. A deformation which satisfies both of these conditions is an *invariant plane strain,* as suggested by Bowles.[115] An invariant plane strain is a strain such that the displacement of all points is in the same direction and proportional to the distance of each point from some fixed reference plane. An example of a two-dimensional invariant plane strain is shown in Fig. 4. The displacement d can be represented as the sum of a simple shear displacement d_1 and a displacement d_2 normal to the reference plane. The illustration can be made three-dimensional by imagining that the deformed square AB'C'D does not lie in the plane containing AD and AB. If the displacement vector points out of the plane of the paper the conditions defining an invariant plane strain are still valid. The displacement of points on the initial line AB to points on AB' are all in the same direction and each of a magnitude proportional to the distance of the initial point from A. Thus, the coordinates of the final positions are linear functions of the co-ordinates of the initial positions; the deformation is *homogeneous* and

$$X_M = F X_P \qquad (1)$$

The subscript M refers to the martensite crystal and P to the parent crystal.

The pure *lattice deformation* changes the lattice of the parent phase into the martensite lattice and consequently it is necessary to assume a unique

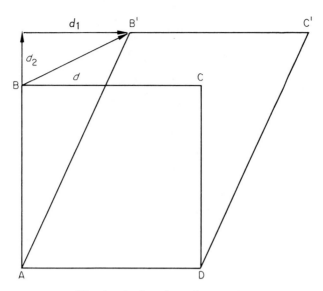

Fig. 4. An invariant plane strain.

correspondence between the two lattices. In general this presents little difficulty, because in most cases it is necessary only to assume that a small volume of the martensite lattice, usually a compound cell, is derived from a volume unit of lattice of the parent containing only one or two unit cells by simple linear strains in the three principal directions. Thus, the deformation can be represented by a set of linear equations

$$X_M = P X_P \tag{2}$$

Such a deformation may not, in all cases, locate the translated atoms exactly on the martensite lattice points. Then small shuffles, necessary to complete the transition, have to be allowed. There is one unusual case in which an intuitive choice of lattice correspondence is not easily made. Lomer[116] examined 1600 possible correspondences between β- and α-uranium–chromium before selecting the most probable one as that involving the minimum displacement. Although there is no formal basis for deciding that the most probable correspondence is that with the minimum strain, this choice has proved satisfactory for many systems which have been examined. The problem of the criterion for selecting correspondence lattices is fundamental, but no clear answer can be given at the present time.

The correspondence for the transformation from face-centred-cubic (f.c.c.) γ-iron to body-centred-cubic (b.c.c.) α-iron proposed by Bain[7] is of historical importance and is the only correspondence for this transformation considered in modern theories. A body-centred-tetragonal (b.c.t.) cell with an axial ratio of $\sqrt{2}$ is selected in the f.c.c. lattice and this is assumed to correspond to a b.c.t. cell with c/a between 1·0 and 1·08 in the martensite crystal (Fig. 5). This correspondence satisfies the minimum

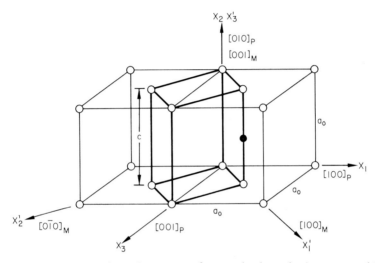

Fig. 5. The correspondence between an f.c.c. and a b.c.t. lattice suggested by Bain.[7] The variant illustrated is that used first by Jaswon and Wheeler.[8]

strain criterion. An essential feature is that one of the principal axes (X_1, X_2, X_3) of the f.c.c. lattice becomes the C-axis (X_3') of the martensite cell so that $\langle 001 \rangle_M$ is parallel to $\langle 001 \rangle_P$. Having selected one of the $\langle 001 \rangle_M$ axes, the other two can be any pair of axes making an orthogonal system with the selected axis. Suitable axes can be found among the $\langle 011 \rangle_P$ directions. In the correspondence illustrated in Fig. 5 the variant with $[010]_P \| [001]_M$ and $[101]_P \| [100]_M$ has been chosen. The correspondence matrices are

$$
\text{M}\underline{\overset{C}{-}}\text{P} = \begin{bmatrix} 1 & 0 & 1 \\ 1 & 0 & -1 \\ 0 & 1 & 0 \end{bmatrix} \tag{3a}
$$

and

$$P\underline{C}M = \tfrac{1}{2}\begin{bmatrix} 1 & 1 & 0 \\ 0 & 0 & 2 \\ 1 & -1 & 0 \end{bmatrix} \tag{3b}$$

Although Bain proposed this specific correspondence the term 'Bain strain' is now used generally to describe the deformation implied by the lattice correspondence in any system.

If the alloy in which an f.c.c.–b.c.c. transformation occurs contains an interstitial solute (for example, iron or iron–nickel containing carbon or nitrogen) the unit cell of the martensite is tetragonal, the tetragonality $\gamma = c/a$ increasing with increasing concentration of solute (Fig. 6). The

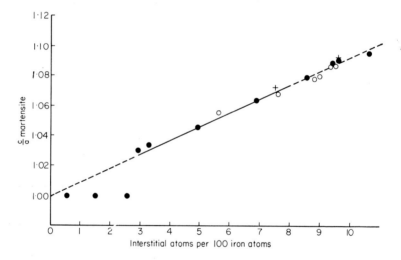

Fig. 6.   The increase in γ with increasing concentration of carbon and nitrogen in iron. The line is the best fit through data for iron–carbon. The points are for iron–nitrogen.[9]

tetragonality is produced by interstitial atoms trapped by the transformation in octahedral sites such as the one marked by a filled circle in Fig. 5. For iron alloys containing carbon or nitrogen γ can be changed only over a small range (1·0 to 1·08 for iron–carbon alloys) and this variation has little affect on the crystallographic parameters such as the indices of the habit plane. If a_0 is the length of the cube edge of the f.c.c. lattice and a

and c are the dimensions of the martensite cell the principal strains on converting the f.c.c to a b.c.c. lattice are

$$\varepsilon_1 = \varepsilon_3 = \frac{\sqrt{2}a}{a_0} \quad \text{and} \quad \varepsilon_2 = \frac{c}{a_0}$$

The next step is to put the lattice deformation (the Bain strain) and the lattice invariant deformation together to produce the total shape deformation and at the same time satisfy the condition that the habit plane is undistorted and unrotated. The latter condition cannot be satisfied solely by the application of the Bain strain to the parent lattice. Consider a unit sphere in the unstrained parent crystal such that

$$X_1{}^2 + X_2{}^2 + X_3{}^2 = 1 \tag{4}$$

If a homogeneous deformation

$$X_M = A X_P \tag{5}$$

is applied, the result can be represented by a strain ellipsoid obtained by substituting

$$X_P = A^{-1} X_M \tag{6}$$

into eqn. (4). This ellipsoid has three orthogonal principal axes and it can be shown that these are derived from three mutually perpendicular axes in the undistorted parent crystal after a rigid-body rotation of the parent. The situation is illustrated in Fig. 7. The strains in the X_1, X_2 and X_3 directions, after rotation of the crystal to bring the directions of the principal strains into coincidence with the orthogonal axes of the parent sphere (eqn. 3), are ε_1, ε_2 and ε_3 and the deformation is pure strain. If the strain is negative in one direction, say X_3, is positive in the X_2 direction, and zero in the third direction, the ellipsoid intersects the sphere in two circles, each with a centre on the axis X_3 (Fig. 7). All lines lying in the surface of a cone containing one of the circles of intersections of the ellipsoid and the sphere and subtending an angle φ_2 at the origin are the same length as the radius of the original sphere. There are two such cones. Before the pure strain is applied the apex angle of the cones is $2\varphi_1$, with

$$\tan \varphi_1 = \left(\frac{1 - \varepsilon_3{}^2}{\varepsilon_2{}^2 - 1} \right)^{\frac{1}{2}}$$

The deformation moves the line BD to B'D' and AC to A'C', the angle changing from $2\varphi_1$ to $2\varphi_2$,

$$\tan \varphi_2 = \frac{\varepsilon_2}{\varepsilon_3} \left(\frac{1 - \varepsilon_3{}^2}{\varepsilon_2{}^2 - 1} \right)^{\frac{1}{2}}$$

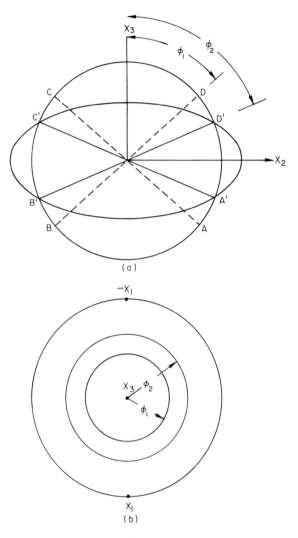

Fig. 7. The Bain cones (a) strain ellipsoid with $\varepsilon_2 > 0$ and $\varepsilon_3 < 0$, (b) stereo-graphic projection of the circles in which the ellipsoid intersects the sphere. The projection is on the plane X_1X_2.

Thus, the Bain strain causes a rotation of the unextended lines, counter-clockwise on the right and clockwise on the left side of the plane X_1X_3. To define an undistorted plane it is necessary to specify two unextended lines which are not mutually parallel and which contain an angle which does not change. In the present situation, an infinite number of planes can be identified which satisfy the first requirement by selecting pairs of lines which lie in the surface of the Bain cones, but none of these pairs can satisfy the second condition. An undistorted plane can be found only by pairing one of the lines in the Bain cone with a line outside of it. A second unextended line which makes an invariant angle with a line in the cone can exist only if one of the principal strains is zero. The complete condition for the presence of an undistorted plane is that one principal strain must be zero, another must be positive and the third must be negative. In general, none of the principal strains of the Bain strain for real systems is zero. In the example illustrated in Fig. 5, ε_1 and ε_2 are small and positive and ε_3 is large and negative. Thus, in general, application of the Bain strain does not leave any plane in the original crystal undistorted. Clearly, the lattice invariant deformation must be such that when it is combined with the Bain strain to produce the total shape deformation there are at least two non-parallel lattice vectors which are unchanged in length and angular relationship. Since an undistorted plane cannot exist when the Bain strain alone is applied, the undistorted plane after the combined Bain and lattice invariant strains have been applied is unlikely to be a rational plane of the parent lattice.

It is assumed that the lattice invariant strain is accomplished by slip or twinning. In the early theories of the crystallography of martensite it was assumed that only one slip or twin mode was operative and consequently the lattice invariant strain was assumed to be a simple shear. This is in agreement with experience for most materials. Usually at any specified temperature and strain rate only one slip or twinning system is operative. An important exception is body-centred-cubic crystals which slip on two or more systems over a wide range of temperature. Complex combinations of slip and twinning systems have been considered, and these will be discussed in a later section of this chapter. For the present, the original assumption of a single operative deformation system will be adopted. If the deformation is a simple shear the plane on which the shear occurs (K_1 if the shear is by twinning) is not distorted or rotated and other undistorted but rotated planes (K_2) can always be found (*see* Fig. 11). It is usually assumed that the simple shear occurs by slip on a single slip system on parallel glide planes or by twinning on parallel planes. When the second deformation is by slip the shear strain is determined by the number and Burgers vector of the dislocations moving on each plane and the spacing between the glide planes. Usually, the density of dislocations generated by the transformation is so great (often in excess of 10^{12} dislocations cm^{-2}) that individual dislocations cannot be resolved by electron microscopy (Fig. 8). The number of dislocations remaining in a single

Fig. 8.   Substructure of martensite in iron–18% nickel alloy. Lattice invariant shear is by slip. Individual dislocations cannot be resolved. Electron micrograph ×60,000 (F. G. Wilson).

martensite plate is far greater than the number of dislocations required to produce the simple shear. In principle, there is no reason why any of the transformation dislocations should remain in the plate, but in a real crystal some will be blocked. If the material has a high stacking-fault energy, some will cross-slip and multiply during glide and possibly some tangling will occur. However, by analogy with the results of deformation of crystals, it seems unlikely that the high density of dislocations observed can be accounted for by these processes alone. An important contributing factor will be discussed later. In materials with a low stacking-fault energy

the transformation dislocations are either generated as extended dislocations or else they very quickly dissociate into partials bounding a ribbon
of stacking fault. In this case the glide dislocations do not leave the glide
plane and the nature of the lattice invariant shear is clearly defined. Arrays
of parallel stacking faults formed by martensitic transformation in copper
alloys have been described in detail by Warlimont.[11]

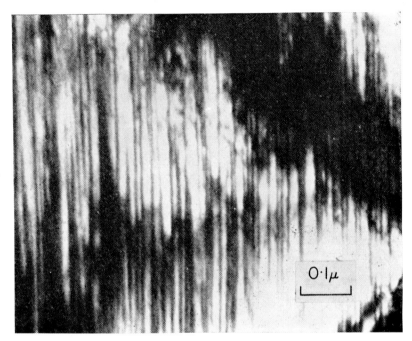

Fig. 9.   Parallel twins in the substructure of iron–30% nickel martensite.
Electron micrograph ×100,000 (Gaggero and Hull[13]).

A number of martensitic transformations are known in which the
lattice invariant shear is produced by twinning. An example is the f.c.c.
to b.c.t. transformation in iron–nickel–carbon alloys with a $\{3, 10, 15\}$
habit in which the lattice invariant strain is produced by twinning on
$\{112\}_M$ planes in a $\langle 11\bar{1} \rangle_M$ direction (Fig. 9). The $\{112\}_M$ plane is derived
from a $\{110\}_P$ plane in the f.c.c. parent lattice. Bowles and Mackenzie[12]
showed that twin planes in the martensite structure are derived from
a mirror plane in the parent crystal. The two variants of the twin should,
in principle, be treated as separate transformation products each with a

total shape deformation F_1 and F_2, the two variants alternating in the stack of thin plates such that

$$F = fF_1 + (1 - f)F_2 \qquad (7)$$

where f is the volume fraction of one variant of the twin orientation. However, this is not necessary, because it is evident from Fig. 10 that the

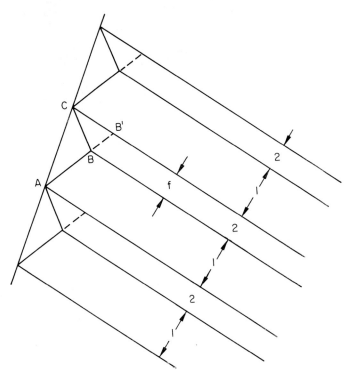

Fig. 10.   Representation of twin variants 1 and 2. The volume fraction of variant 2 is f.

total simple shear produced by the lattice invariant deformation can be considered to be the average shear produced when a fraction f of the crystal is sheared from variant 1 to variant 2. The shear displacement B'C is produced by the twinning of plate 2. The ratio of the thickness of the twinned to the untwinned is $f/(1 - f)$. Then the average shear, which determines the location of the interface AC, is a simple function of the twin shear displacement and the ratio $f/(1 - f)$.

The displacement produced by applying a simple shear to a unit sphere in the unstrained parent crystal can be represented by the strain ellipse in Fig. 11. The plane of the figure is the plane of shear and consequently

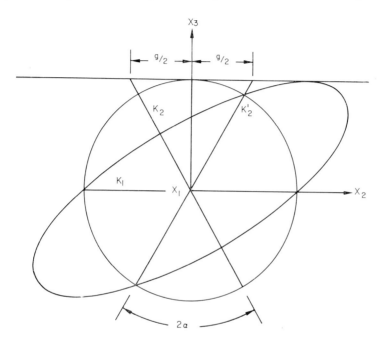

Fig. 11.   The application of a simple shear to a unit sphere in the unstrained parent lattice.

the axes $X_1X_2X_3$ do not necessarily coincide with the axes $X_1X_2X_3$ in Fig. 5. The shear is parallel to the plane X_1X_2 which, adopting the convention used for the analysis of twin displacement, is labeled K_1. The second plane undisturbed by the shear is K_2 which is moved to the position K_2' by the deformation. The lattice invariant strain is

$$X_M = SX_P \tag{8}$$

An example is shown stereographically in Fig. 12. The lattice invariant strain illustrated is a simple shear on $(111)_P$ in the $[\bar{1}2\bar{1}]_P$ direction.

The Bain strain P when combined with the lattice invariant strain S gives the total shape change F. As can be seen in Figs. 7 and 12, this must also involve a rotation R of the lattice to bring the undistorted and unrotated habit plane back into its position in the original lattice. The order

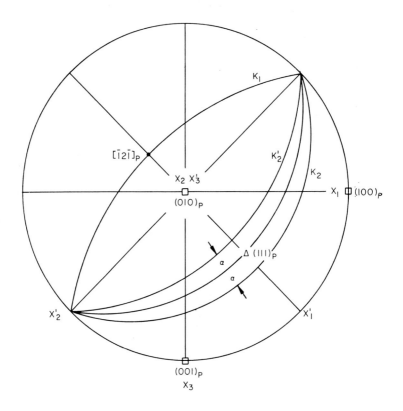

Fig. 12.   Stereogram showing the displacement of the K_2 plane assuming that a simple shear occurs on the (111)$_P$ plane of the f.c.c. structure in the $[\bar{1}1\bar{2}]_P$ direction. It is assumed that the shear angle 2α is 20°. X_1, X_2 and X_3 are the principal axes of the parent lattice and X_1', X_2' are the {001}$_M$ plane normals shown in Fig. 5. (After Lieberman.[14])

in which P, S and R are combined is decided only by the ease with which the algebraic manipulations can be performed. The formal theories describe only the initial and final geometry and do not provide any information about the atom movements by which the individual steps are carried out or the sequence in which these steps occur. In the early Wechsler, Lieberman and Read theories the strain P was applied first and a lattice invariant shear S^M was applied to the product but later theories, including the Bilby–Bullough theory, apply S to the parent lattice and then operate the Bain strain. Using this convention

$$X_M = RPSX_P \qquad (9)$$

The rotation R defines the orientation of the martensite relative to the parent lattice, a parameter which can usually be measured experimentally.

The Bain strain P has to be combined with the lattice invariant strain S in such a way that two non-parallel lattice vectors remain unaltered in length and angular relationship. This is done by a proper selection of the shear angle α (Figs. 11 and 12) and a rigid-body rotation of one lattice.

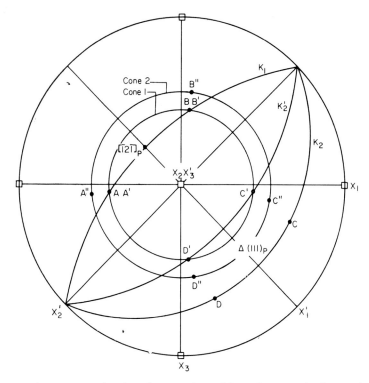

Fig. 13. Stereogram showing the superimposition of a pure lattice strain P and a lattice invariant strain S. (After Lieberman.[14])

This is illustrated by the stereogram in Fig. 13. Following Lieberman,[14] this stereogram is obtained by superimposing the shear displacement due to the lattice invariant strain (Fig. 12) and the projection of the Bain cones (Fig. 7(b)). The example (Fig. 12) used to illustrate the average shear is continued. If the martensite is cubic the semi-apex angles of the cones are $\varphi_1 = 47 \cdot 7°$ and $\varphi_2 = 57 \cdot 2°$, any vector lying in the plane K_1 on which the lattice invariant shear occurs, in this case the plane $(111)_P$, will be undistorted by the shear. All such vectors will be represented by points on the trace $(111)_P$. However, of these only the vectors A and B, which lie in this trace and also in the Bain cone φ_1, will remain unextended after the lattice strain is applied. Thus, it is convenient to define $\overrightarrow{A}$ and $\overrightarrow{B}$ as unit vectors. In the stereogram as it is constructed in Fig. 13, there are two vectors lying in the K_2 plane which are undistorted after the two strains are applied. These are $\overrightarrow{C}$ and $\overrightarrow{D}$ in the parent lattice, which are moved to $\overrightarrow{C'}$ and $\overrightarrow{D'}$ by the lattice invariant shear and to $\overrightarrow{C''}$ and $\overrightarrow{D''}$ by the Bain strain. $\overrightarrow{C'}$ and $\overrightarrow{D'}$ lie at the intersection of the trace of the plane K'_2 and the Bain cone φ_1. Thus, four unextended vectors have been identified. It remains now to select values of α such that the angle between pairs of unextended vectors is unchanged by the two strains. For example, if the lattice invariant shear is such that the angle between $\overrightarrow{A}$ and $\overrightarrow{C}$ is equal to the angle between $\overrightarrow{A''}$ and $\overrightarrow{C''}$ then the plane defined by $\overrightarrow{A}$ and $\overrightarrow{C}$ will be undistorted by the transformation. It will, however, be rotated. The angle through which it is rotated is the angle through which $\overrightarrow{A''}$ and $\overrightarrow{C''}$ must be rotated to make them coincide with $\overrightarrow{A}$ and $\overrightarrow{C}$ respectively. This angle, which describes the orientation of the martensite relative to the parent crystal, can be determined easily from the stereogram. In the example under discussion, the value of the angle α which ensures that a constant angle between $\overrightarrow{A}$ and $\overrightarrow{C}$ is maintained also results in the angle between $\overrightarrow{B}$ and $\overrightarrow{D}$ being unchanged by the transformation and, consequently, in this example the solution is degenerate.

Lieberman[14] has described a method for finding values of α for which pairs of unextended vectors with an unchanged angular relationship exist by systematic trials on a stereogram. For example considered in Figs. 12 and 13, there are two values of α: α_1, the shear for which there is no change in angle between $\vec{B}$ and $\vec{C}$, and α_2 for no angular change between $\vec{A}$ and $\vec{C}$. $\vec{A}$ and $\vec{B}$, which lie at the intersection of Bain cone with semi-apex angle φ_1 and the trace of the plane on which the lattice invariant shear occurs, are independent of the extent of the lattice invariant shear. In general, the plane on which the shear occurs is not known. It is usual to assume that it is a plane of one of the known deformation modes. In the example it is assumed that the plane is $(111)_p$ because this is the plane on which both slip and twinning occur in the f.c.c. lattice. The chosen direction, $[\bar{1}2\bar{1}]_p$, is a twinning direction. Of course, as will be discussed later, other slip or twinning modes could have been selected. The vectors $\vec{C}$ and $\vec{D}$ change with α and there are two vectors corresponding to $\vec{C}$ which do not change in length when the lattice invariant shear and the Bain strain are applied and which make the same angle with B before and after the strains are applied. These are $\vec{C}_1$ and $\vec{C}_2$ corresponding to α_1 and α_2, respectively (Fig. 14). Similarly, there are two $\vec{D}$ vectors; $\vec{D}_1$ and $\vec{D}_2$. Consequently, there are four undistorted planes; the planes defined by $\vec{A}$ and $\vec{C}_1$, $\vec{A}$ and $\vec{C}_2$, $\vec{B}$ and $\vec{C}_1$, and $\vec{B}$ and $\vec{C}_2$. Stereographically one of the planes can be found by drawing tne trace of a plane through $\vec{A}$ and $\vec{C}_2$ and finding the pole N_2. Each of the four undistorted planes has a pole N, arbitrarily N_1 ($\vec{A}$ and $\vec{C}_1$), N_2 ($\vec{A}$ and $\vec{C}_2$), N_3 ($\vec{B}$ and $\vec{C}_1$) and N_4 ($\vec{B}$ and $\vec{C}_2$). As noted earlier the other possible undistorted planes, defined by $\vec{B}$ and $\vec{D}_1$, $\vec{B}$ and $\vec{D}_2$, $\vec{A}$ and $\vec{D}_1$ and $\vec{A}$ and $\vec{D}_2$, crystallographically are equivalent to N_1, N_2, N_3, and N_4 respectively, but this is a consequence of symmetry in the particular example under discussion. In general, each of the solutions is independent. The conditions for degeneracy have been discussed by Christian[15] and Wechsler.[16] For some purposes greater accuracy than can be obtained by graphical methods is necessary and then the techniques of matrix algebra must be used. These have been summarised recently by Christian[3] and Wayman.[4] It is possible to determine values of α stereographically to within 1% of the values calculated by matrix techniques.

For many transformations the lattice invariant shear is not known. This is the case, for example, when the shear is by slip and the dislocations cross-slip out of the glide plane so that the plane cannot be identified by conventional electron microscopy. Thus, the crystallographic problem as commonly posed consists of finding a simple shear, usually selected from

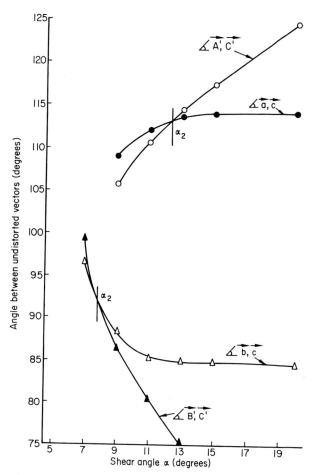

Fig. 14. The angle between unextended vectors as a function of the shear angle α. Two values of α for which there is no angular distortion of the unextended vectors are indicated. (Lieberman.[14])

known deformation modes, which will give the observed crystallographic parameters, it being assumed that the dimensions of the unit cell of each lattice and the lattice correspondence are known. The lattice invariant shear is defined by two direction cosines of the normal to the shear plane and the direction and magnitude of the shear. The measurements against which the predictions can be tested are the habit plane (two direction cosines of the normal), the orientation (two direction cosines of the axis of rotation and the angle of rotation) and the total shape change (two direction cosines of the displacement and its magnitude). Thus eight quantities can be measured and the problem is overdetermined. However, all the measurements cannot be made with the same accuracy and usually comparison of predicted and measured habit planes is the first test to be applied. In many systems the orientation is not very sensitive. to changes in the shear selected, and only recently have sufficiently accurate measurements of the shape change been made to warrant the use of these parameters for comparison with the predictions. Recent work by Wayman's group and others has underlined the importance of comparing the theoretical predictions with all of the parameters which can be measured.

The orientation between the parent phase and the martensite crystal is obtained, by either the stereographic or the algebraic method, in the form of an axis and an angle of rotation. This is the least ambiguous method of representation but it is common practice to express the orientation as the deviation from parallelism of a close-packed plane in the martensitic and parent structures. The Bain strain is homogenous and the correspondence is usually between relatively simple cells; consequently, when the Bain strain alone is applied it is usual to find certain rational planes and directions in the product parallel to rational planes and directions in the parent lattice. These relationships are not affected by the lattice invariant strain but when the rotation is applied the simple parallels disappear. However, it is usually possible to find some rational planes and directions which are nearly parallel and these are used to define the orientation. Two well-known orientations found experimentally for martensites in iron alloys are the Kurdjumov and Sachs[17] orientation.

$$(111)_P \| (011)_M$$

$$[\bar{1}10]_P \| [\bar{1}1\bar{1}]_M \tag{10}$$

and that found by Nishiyama[18]

$$(111)_P \| (110)_M$$

$$[\bar{2}11]_P \| [1\bar{1}0]_M$$

These differ only by a rotation of $5° \ 16'$ about $[111]_P$.

3. ACICULAR-MARTENSITE

The theories of the crystallography of martensite involve a number of assumptions. In the simplest form, as set out by Wechsler, Lieberman and

Read,[19] the most important assumptions are that the interface is undistorted, which is equivalent to assuming that the strain energy associated with the interface is a minimum, that the lattice invariant deformation is a simple shear which is produced by a single known deformation mechanism and that the lattice correspondence is such that at least one major axis of the product lattice coincides with a major axis in the parent lattice. Some alloy systems are known in which the martensitic transformation is simple and the predictions of the crystallographic theories are found to be in excellent agreement with the observed crystallographic elements. Examples are the transformations in gold–cadmium and in iron–platinum. However, for most transformations significant, and in some cases large, discrepancies have been found between the theoretical predictions and the experimental measurements. Nevertheless, the theory has proved to be an invaluable frame of reference within which to consider non-ideal behaviour in real systems and useful ideas about the possible nature of the deformation accompanying transformations have evolved from a detailed consideration of the deviations.

The martensitic transformation in gold–cadmium alloys, from a parent phase which is ordered b.c.c. to a product which is ordered orthorhombic, has received a lot of attention because it appears to represent a nearly ideal system. The lattice correspondence is assumed to be

$$
_{\mathrm{M}}\underline{C}_{\mathrm{P}} = \begin{bmatrix} 1 & 0 & 0 \\ 0 & \frac{1}{2} & -\frac{1}{2} \\ 0 & \frac{1}{2} & \frac{1}{2} \end{bmatrix}, \quad _{\mathrm{P}}\underline{C}_{\mathrm{M}} = \begin{bmatrix} 1 & 0 & 0 \\ 0 & 1 & 1 \\ 0 & -1 & 1 \end{bmatrix} \tag{11}
$$

The most complete treatment of the theory applied to this transformation is that given by Mackenzie and Bowles.[20] Accurate measurements of the habit plane and the orientation in a partially transformed crystal containing 47·5% of Cd have been made by Lieberman, Wechsler and Read.[21] The parent can be transformed by the movement of a single interface, thus eliminating errors due to accommodation strains. The temperature of transformation (M_s) is about 60°C and the movements of the interface can be controlled by control of the temperature gradient in a single crystal. The parent and product lattices are ordered and this provides a further restriction on the selection of possible lattice correspondence. It is observed that the internal structure of the martensite plates consists of a stack of twins on $\{111\}_{\mathrm{M}}$ planes, indicating that the lattice invariant shear is a simple shear of the kind discussed earlier. An important feature of the Bowles–Mackenzie theory is the assumption that the plane of the lattice invariant shear is a mirror plane of the parent lattice which becomes a twin plane in the martensite. Thus, they assumed a shear occurs on $(1\bar{1}0)$ of the parent, which corresponds to the twin plane $(1\bar{1}\bar{1})$ in the martensite. The twin direction in the martensite is irrational. The predicted habit plane is $(3\bar{1}3)_P$ which is within $1\cdot5°$ of the observed habit. The orientation was found to be $(001)_M\|(011)_P$ and $[110]_M\|[11\bar{1}]_P$ and for both planes and directions the theoretical predictions agree with these experimental data within the limits of accuracy of the measurements. The observed shape change is also close to that predicted. Thus, overall, the agreement between theory and observation is good. However, it must be emphasised that this is an unusually simple transformation in which the matching of the lattices across the interface must be nearly exact as shown by the mechanical reversibility of the transformation (Chang and Read[22]) and the fact that the good agreement between the theoretical and experimental crystallographic parameters is obtained without the introduction of a dilatation parameter.

Another particularly simple transformation is the f.c.c. to f.c.t. transformation in indium–thallium alloys. Burkhart and Read[23] studied an alloy containing $20\cdot7\%$ thallium. The habit plane predicted by the theory is less than $0\cdot5°$ from $\{011\}_P$ which is the habit found experimentally. This transformation is notable because of the extremely small Bain strains involved.

The only transformation in an iron alloy which is known to be nearly ideal is that reported recently by Efsic and Wayman[24] in iron–$24\cdot5$ a/o platinum. Isolated undistorted plates of martensite were obtained and the habit plane, orientation relationship, and the magnitude and direction of the shape deformation were all measured with good accuracy. The agreement with the theoretically predicted parameters was, in the words of the authors, 'essentially perfect'. It is particularly notable that in this alloy, as in the gold–cadmium and indium–thallium alloys, the deformation in the parent phase adjacent to a martensite plate is too small to be detected (Fig. 15). The plastic strains accompanying the transformation are zero and the elastic strains must be small.

The shortcomings of the simple theory can be illustrated by considering the martensite transformation in iron–nickel–carbon alloys. Early work on iron–carbon alloys revealed that when the carbon is between about $0\cdot4\%$ and $1\cdot4\%$ the habit plane is $\{225\}_P$ and the orientation is Kurdjumov–Sachs, but alloys with carbon between $1\cdot4\%$ and $1\cdot8\%$ transform to martensite with a $\{259\}_P$ habit and a Nishiyama orientation. Binary iron–nickel alloys with between $28\cdot5\%$ and 33% nickel also transform to

Fig. 15.   Interferogram of a martensite plate formed in an iron–24·5% platinum
alloy, ×245. (Efsic and Wayman.[24])

$\{259\}_P$ martensite with an orientation that is close to Nishiyama. Iron–nickel alloys with less than 28·5% nickel and dilute iron–carbon alloys (with less than about 0·4% carbon) transform to martensite with a habit near to $\{111\}_P$,[25] and this will be discussed in the next section. The iron–carbon martensites with $\{225\}_P$ or $\{259\}_P$ habit are tetragonal, $\gamma = c/a$ increasing from 1·0 to 1·08 as the carbon is increased up to 1·8%. Even when cooled to 4°K the transformation is incomplete so that in all specimens untransformed parent phase (retained austenite) coexists with the martensite plates. Kelly and Nutting[26] showed by electron transmission microscopy that the substructure of the martensite plates consists of stacks of twins on $\{112\}_M$, but the regularly twinned structure often extends only part way through the thickness of the plate, although it is usually found on both sides of the midrib. In the outer regions the twins are usually replaced by a dense aggregation of dislocations. The trace of the midrib is straight but the interface between the martensite and retained austenite is often irregular, especially when the volume fraction of martensite containing a dislocation substructure is large. Patterson and Wayman[27] suggested that the change from a twinned to a dislocated substructure occurs because the heat of transformation increases the temperature of the lattice near to the moving austenite–martensite interface and, at a certain stage in the growth, the temperature becomes too high for continued growth of the twinned form of martensite. If the austenite–martensite interface of a fully grown plate is assumed to be the habit plane, the scatter in the experimental values is large and the appropriate lattice invariant shear is unclear. However, even when the habit is taken to be the plane of the midrib, the scatter in the measurements on one sample is large and is greater than the estimated experimental error (Otte and Read[28]).

It is reasonable to assume that the lattice invariant shear in the iron–nickel–carbon martensites with a habit near $\{259\}_P$ is a twin shear on $\{112\}_M$. The usual twin direction in a b.c.c. lattice is $\langle 111 \rangle_M$ and this is usually assumed to be the direction of twinning in a martensite plate also. The assumed Bain correspondence (*see* Fig. 5) is such that $\{112\}_M \langle 111 \rangle_M$ is derived from $\{110\}_P \langle 110 \rangle_P$. This was the simple shear assumed in the original Wechsler–Lieberman–Read theory. Instead of the plane of shear being the $(111)_P$ glide plane as in the example in Fig. 12, the shear was assumed to occur on $(011)_P$ in the $[011]_P$ direction. The first major success of the Wechsler–Lieberman–Read theory was in predicting habits in iron–nickel–carbon martensite close to $(3, 10, 15)_P$. Greninger and Troiano[5] accurately determined the habit (given as between $(3, 10, 15)$ and $(9, 22, 33)$) and orientation of the martensite plates in an iron–22% nickel–0·08% carbon alloy. The habit and orientation predicted by Wechsler–Lieberman–Read assuming shear on $(011)_P [01\bar{1}]_P$ and $\alpha = 7°$ implies that the ratio of the thickness of alternate twin lamellae is about 2:1. The Wechsler–Lieberman–Read theory does not predict the $\{225\}_P$ habit found in other alloys if the lattice invariant shear is assumed to be twinning on $\{112\}_M$. However, a habit not far away from $\{225\}_P$ is predicted if the lattice invariant shear is slip on $\{111\}_P$. This is the shear considered in the example (Fig. 12).

Accurate measurements of the orientation between a martensite plate and the surrounding austenite are difficult to make and reported results are scarce. However, there are sufficient experimental data available to show that as the habit changes significantly (say, from $\{225\}_P$ to $\{259\}_P$) the orientation changes only a few degrees. Thus, orientation is a relatively insensitive parameter and most attention has been focused on the measurements of the habit plane. Experimental work since the studies of Greninger and Troiano (Johnson and Wayman[29] and Otte[30]), have included measurements on iron–chromium–carbon alloys as well as on an extended range of iron–nickel–carbon alloys and has shown that $\{225\}_P$ and $\{259\}_P$ are not discrete habits but that the poles of habit planes are scattered over large areas of the stereographic triangle. Some of the data, compiled by Crocker and Bilby,[31] are shown in Fig. 16. A number of possible causes of this wide variation have been suggested and examined; variation in the Bain strain due to changes in the degree of tetragonality,[25] the introduction of a distortion at the interface in the form of a uniform dilation[12] (quite recently anisotropic interfacial distortions have been considered), and the operation of unusual or compound modes of deformation to produce the lattice invariant shear.[31] The tetragonality γ can be changed by about 8% but this produces only a small change in the habit plane normal and is quite inadequate to explain the large variations observed. The possibility of an isotropic distortion in the habit plane was considered by Bowles and Mackenzie[12] who introduced a scalar parameter δ such that the principal strains in the matrix became $\delta\varepsilon_i$. Since δ is usually considered to be greater than unity, this has the effect of reducing the apex angle $2\varphi_2$ of the Bain cone (*see* Fig. 7). The upper limit to the value of δ is set by the requirement that stereographically the trace of the plane on which the invariant shear occurs must intersect the trace of the Bain cone (*see* Fig. 13). If the shear is by twinning so that the plane involved is $\{011\}_P$ and γ is taken at its maximum value of 1·08, the upper limit of δ is 1·02. The extent to which the predicted habit plane is changed is plotted in the stereographic triangle (Fig. 17). At a constant value of γ, the habit plane moves continuously along the curve as δ is varied between the limiting values. At the largest value of δ the habit is close to $(225)_P$. Only a small value of δ (1·002) is required to improve the already good agreement between the Wechsler–Lieberman–Read theory and observed Greninger–Troiano habit (3, 10, 15)$_P$. The simple theory fails completely when applied to the $(225)_P$ habit but the main features of this transformation can be explained satisfactorily by the introduction of the Bowles–Mackenzie dilatation parameter δ. However, a value of δ near to the maximum possible is required and the physical significance of the δ parameter is unclear. As conceived by Bowles–Mackenzie it describes a uniform expansion of the martensite lattice to balance the $1 \cdot 8 \%$ contraction of the close-packed $[1\bar{1}0]_P$ line into $[111]_M$ during transformation. The artificial assumption is made that this local elastic accommodation extends only a few atomic distances into the lattice. The validity of the concept of a δ parameter is currently the subject of a

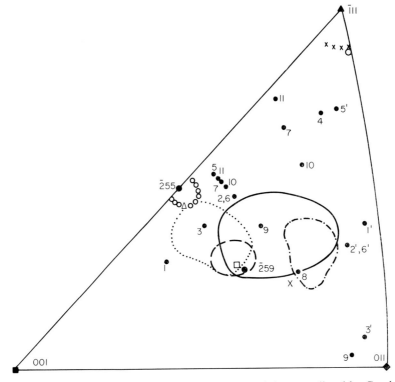

Fig. 16.   Comparison of measured habit planes and those predicted by Crocker and Bilby.[31] Regions of scatter of experimental habit plane results: ○ ○ ○ ○ Iron–0·95% carbon, iron–1·4% carbon (Greninger and Troiano[49]). – – – – Iron–1·78% carbon (Greninger and Troiano[49]). — · — · — Iron–32·5% nickel (Greninger and Troiano[49]). ——— Iron–30% nickel (Machlin and Cohen[113]). ...... Iron–2·8% chromium–1·5% carbon (Otte and Read[28]). x x x x Pure iron (Entwisle[114]). X Iron–0·8% carbon–22% nickel (Greninger and Troiano[5]). Predicted habit planes: ● 1–4 (111)$_P$; $[1\bar{1}0]_P$, $[10\bar{1}]_P$, $[\bar{2}11]_P$, $[11\bar{2}]_P$ respectively. 5–11 $[111]_M$; $(1\bar{1}0)_M$; $(10\bar{1})_M$, $(\bar{2}11)_M$, $(11\bar{2})_M$, $(12\bar{3})_M$, $(\bar{3}12)_M$, $(2\bar{3}1)_M$ respectively. ○ Near $\{111\}_P$ pole predicted by equal amounts of shear on $(1\bar{1}0)_M$ $[111]_M$ and $(1\bar{1}0)_M$ $[11\bar{1}]_M$. □ Near $\{259\}_P$ pole predicted by shear on $(\bar{3}42)_P$ $[211]_P$. Δ Near $\{225\}_P$ pole predicted by shear on $(\bar{1}\bar{1}, 10, 12)_P$ $[211]_P$.

major controversy. It is improbable that the matching conditions at the interface can be satisfied by an anisotropic strain or that strains as large as 2% can be introduced elastically. Furthermore, Krauklis and Bowles[117] have shown recently that for a particular $\{225\}_P$ transformation no change in the length of the close-packed line occurs. Thus, the macroscopic change in lattice spacing must be accommodated in some other way.

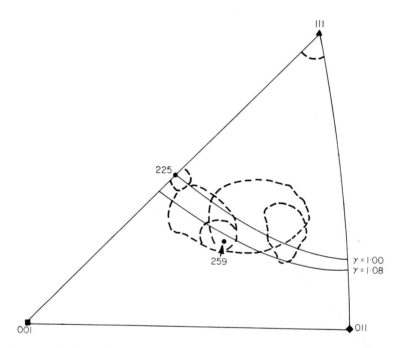

Fig. 17.   The habit planes predicted by the formal theories lie along a smooth curve when the dilatation parameter δ is varied. The curves for the limiting axial ratios $\gamma = 1\cdot00$ and $\gamma = 1\cdot08$ are shown and compared with the experimental data. The observed habit planes lie within the areas enclosed by the broken lines. These experimental results are described more fully in Fig. 16.

Crocker and Bilby[31] set up two computer programs to reduce the labour involved in obtaining numerical solutions from the formal crystal-lographic theories; one program gave the habit plane from the Bain strain and the plane and direction of the lattice invariant strain and the other gave the orientation relationship from the same input data. These programs were used in a study designed to test the extent to which the observed habit planes can be predicted if more general shear systems than those used by Wechsler–Lieberman–Read and Bowles–Mackenzie are allowed. They examined data relating to more than 3000 different habit planes. Their results, assuming no interface dilatation, will be described by reference to Fig. 16. The poles of the habit planes predicted using normal deformation modes of the austenite or martensite ($(\bar{1}1\bar{1})_P$ with $[\bar{1}01]_P$, $[\bar{1}\bar{1}0]_P$, $[21\bar{1}]_P$ or $[\bar{1}2\bar{1}]_P$ and $[111]_M$ shear on $(1\bar{1}0)_M$, $(10\bar{1})_M$, $(\bar{2}11)_M$, $(11\bar{2})_M$, $(12\bar{3})_M$, $(\bar{3}12)_M$, or $(2\bar{3}1)_M$) for the lattice invariant shear all plotted outside of the areas of the sterographic triangle containing the poles of habit planes observed experimentally, except for the poles 8, 3 and 9. Pole 8 corresponds to the (3, 10, 15) habit predicted by Wechsler–Lieberman–Read, the lattice invariant strain being $(11\bar{2})_M [111]_M$. Pole 3 corresponds to twinning of austenite and pole 9 to slip of martensite on $\{123\}_M$, but neither of these deformation modes is commonly observed experimentally. More general systems involving either shear in a fixed direction on two different planes, or on a selected plane in two directions, were also tried. One such system can be used to explain the $\{111\}_P$ habit which will be referred to later, but the overall shape change is large. In general, this type of shear dissociation does not significantly improve the agreements between theory and observation and the situation is not improved if a small dilatation ($\delta <$ 1·005) is allowed.

With the recent experimental evidence that more than one lattice invariant shear mechanism is operative in Fe–C[118] and Fe–Ni[119] alloys, the possibilities of explaining some of the previously unexplainable crystallographic data by a multiple-shear lattice-invariant deformation became realistic. Acton and Bevis,[120] Ross and Crocker,[121] and Dunne and Bowles[122] have formulated crystallographic theories which allow the possibility of two or more rational inhomogeneous shear modes occurring simultaneously. In the Acton–Bevis and Ross–Crocker theories the relative amounts of shear on the specified systems are permitted to vary continuously subject only to the condition that the overall shape strain is an invariant plane strain. The Bowles–Dunne theory is more specialised in that it specifies one of the inhomogeneous shears to be due to accommodating plastic flow in the austenite. It now appears that these theories will be capable of explaining virtually all of the crystallographic data on acicular-martensite.

When the lattice-invariant deformation is produced by multiple shears the dislocation structure of the interface between the martensite and parent crystal must be complex. There is an additional requirement. These interfacial dislocations must be glissile because it is by the movement of this interface that the total transformation, including the lattice invariant deformation, is produced. Study of the problem in these terms is underway at the present time.

4.   MASSIVE-MARTENSITE

A martensitic structure with a morphology completely different from that of acicular-martensite described in the preceding section was first identified about ten years ago. Plates in pure iron and in iron-nickel alloys with a $\{111\}_P$ habit were reported as long ago as 1929.[32,33,34] It was known that

Fig. 18.   Appearance of massive-martensite on a transverse section through an iron–25% nickel specimen. Etched in ferric chloride. ×350.

the morphology of the structure of quenched low-carbon ($<0.4\%$ carbon) iron–carbon alloys is different from that of high-carbon alloys but the nature and importance of these differences was not realised. Kelly and Nutting[26] carried out a detailed study by electron transmission microscopy of martensite in an iron–0.1% carbon alloy and showed that the substructure consists of 'needles' or 'laths' about 1μ in diameter aligned with the major axes parallel to $\langle 011 \rangle_p$. Each lath contains a high density of dislocations but no evidence of twins. Surprisingly, the morphology on a more macroscopic scale of martensite in dilute iron–carbon alloys was not described systematically and in detail until the work of Marder and Krauss in 1967.[35] Gilbert and Owen[36] recognised that the martensitic structure in quenched iron–14 a/o nickel alloys does not have the acicular appearance of martensite in higher nickel (28·5–33%) alloys when viewed on a polished

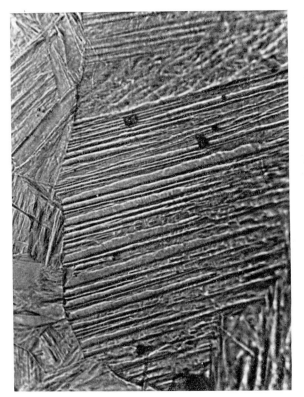

Fig. 19.   Surface relief on iron–20% nickel alloy fully transformed to massive-martensite. Direct illumination. ×480.

and etched transverse section, the transformed structure occupying blocks of irregular shape but with many plane interfaces (Fig. 18). It was soon realised[25] that each block contains many parallel shear plates (Fig. 19) and that the substructure of this type of martensite is the same as that reported earlier[26] for low-carbon steels. By 1964 it was possible to give a complete qualitative description of this form of martensite and to list some of the many alloy systems in which it occurs (Owen, Wilson and Bell[25]); nearly pure iron, iron–carbon ($<0.4\%$ carbon), iron–nitrogen ($<0.4\%$ nitrogen), iron–nickel ($<28.5\%$ nickel), and other substitutional binary alloys of iron such as iron–silicon, iron–copper and iron–manganese. The transformation is from f.c.c. austenite to b.c.c. martensite. All these transformations occur at high temperatures compared with those at which acicular-martensite forms and all are complete at temperatures above 20°C. No austenite is retained in the structure at room temperature. It is difficult to distinguish the individual lamellae (Fig. 19) by etching a polished surface although this has been done recently (Fig. 18). The 'laths' seen by electron microscopy are probably the same as the shear lamellae observed by optical interferometry of surfaces which were flat before transformation, because the shape and dimensions are closely similar, but a direct correspondence has not been demonstrated.

It has proved to be unusually difficult to obtain quantitative information about the structural features of massive-martensite. Interference patterns from the surface of transformed acicular-martensite show that the surface of the sheared plate is flat and the apex is sharp but the patterns from massive-martensite show that the shear angle varies greatly and the ridges are rounded and irregular (Fig. 20). Possibly the ridges are sharp when first formed but quickly blunted by surface diffusion at the relatively high transformation temperatures, but a more probable explanation is that each shear lamella forms in a plastically distorted parent lattice and is further distorted when the next adjacent plate grows. Coexisting untransformed parent phase and the martensitic product are necessary if unambiguous measurements of the crystallographic parameters are to be obtained. With the help of Nilles[135] we have solved this problem by two- and three-surface analysis of partially transformed single crystals of iron–nickel alloys. Isolated plates of massive martensite were obtained by arranging for a pronounced gradient of nickel concentration within the crystal. Careful study of these isolated plates[124,125] (Fig. 21) has shown unambiguously that the shear lamellae observed in the surface topology experiments are traces of plates stacked parallel to each other. The habit plane of the plates of massive-martensite is near to $\{111\}_P$.[125]

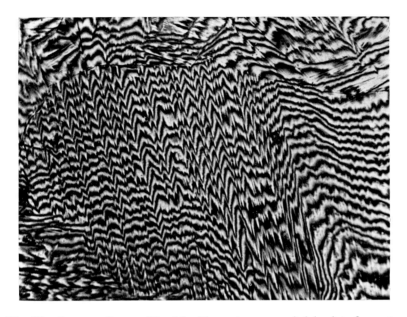

Fig. 20.   Same surface as Fig. 19 with contours revealed by interferometry.
×480.

There have been reports that some of the 'laths' seen by electron transmission are twin related (Speich and Swann[38]), and it has been suggested
that these relationships can be explained if each lath is a variant of the
Kurdjumov–Sachs orientation. However, electron diffraction studies by
other observers have failed to confirm the occurrence of twin-related
plates and there is no direct evidence that the orientation relationship
between the parent-phase and massive-martensite is, in fact, Kurdjumov–
Sachs. Using the partially transformed specimens mentioned earlier and a
special X-ray technique we have determined[124] the orientation relationship in iron–24 nickel alloys. The orientation is found to be close to
Nishiyama. Thus the proposal of Speich and Swann is untenable.

A habit close to $\{111\}_P$ cannot be predicted by the formal theories if
the lattice invariant shear is a single normal deformation mode even if
extreme values of δ are used (*see* Fig. 17). It has been suggested (Wayman[4])
that the surface shears seen by optical microscopy are traces of needles, or
laths, lying parallel to a $\langle 101 \rangle_P$ direction. Such needles could be formed
by degeneracy of $\{225\}_P$ plates but it is not clear how or why such degeneracy should occur. Most of the proponents of the needle or lath concept

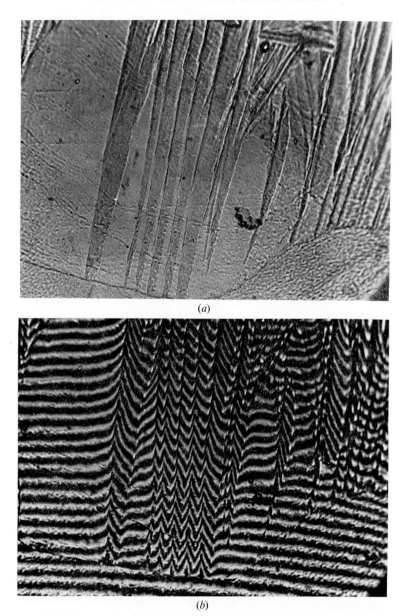

(a)

(b)

Fig. 21. Isolated plates of massive-martensite embedded in the parent austenite. Iron–nickel alloy, (a) direct illumination of surface, ×480; (b) interferogram, ×480. Note the large distortion of the austenite adjacent to the martensite plates.

rely on observations by electron transmission through thin foils, which is a much less satisfactory way of determining the shape of a three-dimensional solid than three-surface analysis by optical microscopy. In view of the work at Cornell mentioned earlier,[37,124,125] it must be accepted that the martensite crystals are plates with a habit close to $\{111\}_P$. Crocker and Bilby's exploration[31] of possible lattice invariant shear modes revealed that a $\{111\}_P$ habit could be predicted if the invariant shear is $(1\bar{1}0)_M$ $[110]_M$ or equal displacement on $(110)_M$ in the $[111]_M$ and $[11\bar{1}]_M$ directions. This is shown by the open circle in Fig. 16. In all alloys which transform to massive-martensite the product is cubic so that $\gamma = 1$ and in the Crocker–Bilby analysis it was assumed that $\delta = 1$ also. The shear $(1\bar{1}0)_M$ $[110]_M$ is rational but unusual. The orientation predicted is substantially different from Nishiyama and is, in fact, closer to Kurdjumov–Sachs. The predicted shear angle is large. Unfortunately, a great deal of cross-slip and tangling of the very numerous dislocations in the martensitic plates makes it impossible to identify the transformation dislocations and consequently the Burgers vectors have not been determined nor has $\{110\}_M$ $\langle 110 \rangle_M$ slip been identified in martensite by trace analysis or any other technique.

The Crocker–Bilby solution adequately explains the habit but fails to predict correctly other important parameters. A solution by Wechsler, Read and Lieberman explains the orientation relationship and the assumed magnitude of the overall shear angle[123] but gives a habit plane about $7\frac{1}{2}°$ from $\{111\}_P$. The burgers vector of the transformation dislocation may, in this prediction, be dissociated into two low energy $\langle 111 \rangle_M$ types. While some of the features of massive-martensite crystallography can be explained by either the Crocker–Bilby or the Wechsler–Read–Lieberman suggestions, neither is completely satisfactory. An analysis using the multiple-shear concept proposed by Acton and Bevis[120] is underway at the present time. Preliminary results are promising, but it is evident that a satisfactory understanding of all aspects of $\{111\}_P$ martensite must await the conclusion of further experimental work in addition to this and other formal analyses.

It is now clear that the essential difference between acicular and massive-martensite is in the extent of the plastic deformation of the parent lattice which accompanies the formation of each plate. If the distortion in the habit plane and the shape change is very small, single interface transformation occurs. With a somewhat larger shape change but still very little distortion, acicular-martensite is formed exactly as predicted by the formal theories of the crystallography (as in iron–platinum). Even with some distortion in the habit plane, represented by a dilatation parameter or some other adjustment, acicular-martensite forms but the correct method of introducing the distortion into the theory is not always clear. Massive-martensite is different; the shape change is certainly accompanied by extensive plastic deformation of the parent and previously formed martensite lattices near to the habit plane. The extensive self-accommodation is probably the reason why the transformation to massive-martensite in chemically homogeneous crystals goes to completion. Consequently, it is not surprising that formal theories assuming one or two simple shears are inadequate and it is for this reason that analyses involving multiple lattice-invariant shears are being developed.

5.   RELAXATION OF CONSTRAINTS

Many martensitic structures which are variants of the structures formed in bulk crystals have been reported. All have formed under conditions in which the elastic constraints imposed on a growing martensite plate by the surrounding parent crystal have been partially relaxed. The effect of elastic, and possibly plastic, constraint can be illustrated by comparing martensite crystals formed in two bulk specimens which differ only in that one is a single crystal and the other a polycrystalline aggregate of the parent phase. The martensite plates formed in the single crystal are usually thin but quite flat. Those formed in the polycrystalline specimen are confined to a single crystal within the aggregate and are lenticular, the plates coming to a point at the boundary to minimise the accommodation shear required near the interface with a neighbouring untransformed crystal.

Spontaneous transformation during the thinning of foils for examination in the electron microscope has been observed in β-brass and in iron–30% nickel (Warlimont,[39] Hull[40] and Gaggero and Hull[13]). These transformations occur only in the thinnest section of the foil (less than 1000 Å) where the constraints are minimal, and at temperatures significantly higher than the transformation temperature (M_s temperature) of the bulk specimens. The structure of the product of these spontaneous transformations is similar to that of the product of the martensitic transformation in the bulk specimen, although some differences have been noted (Hull[40]). The deformation producing the lattice invariant strain is the same as in the bulk transformation. For example, in iron–30% nickel the shear is by twinning (Nishiyama and Shimizu[41]) and the same parallel twins are

found in the transformation product in the thin foil. However, the martensite in the thin foils does not form plates but it is in irregular shaped masses.
The fact that the interface between the parent phase and the martensite is
not planar is difficult to explain in terms of the formalism of the crystallographic theories, even assuming that the constraints are negligible. The
interface might be made up of many small flat surfaces.

Another form of martensite which forms only because some of the
constraint imposed by the parent lattice is removed is surface martensite.
This has been studied most extensively in iron–29·5% nickel alloys. Long,
thin needles, which appear to lie in $\{112\}_P$ planes, form spontaneously at

Fig. 22. Interferogram of surface-martensite on an iron–28% nickel specimen.
$\times 600$.

room temperature on a polished surface of this alloy. More needles can be induced to form by scratching the surface and different complex patterns of intersecting needles can be formed if, during the preparation of the surfaces, different forms of abrasion are applied. The interferogram (Fig. 22) demonstrates that the long, thin volume of martensite has formed by shear and it has been shown by two-surface analysis that the needles grow only to a depth of a few microns.[42] It is difficult to define a habit plane of a needle with a small cross-section. Klostermann[43] believes that the habit is $\{112\}_P$ but that there is no invariant plane strain of the kind required by the crystallographic theories. The orientation relationship was found to be the 'standard variant' lying between the Kurdjumov–Sachs and the Nishiyama orientations. He suggested that this orientation allows $\{112\}_P$ to transform to $\{123\}_M$ without rotation and with perfect atomic matching across the interface. Presumably this definition of the growth conditions is possible because the strain energy associated with a needle or plate growing on a free surface is small.

6. BAINITE

The term 'bainite' is usually applied to complex transformations which involve both a shape change and diffusion; the solvent lattice transforms to produce a shape change[133,134] similar to that in a martensitic transformation but the solute atoms, usually interstitial, diffuse over an appreciable lattice distance. In general, the diffusion is directed towards a growing particle of a precipitate phase, although perhaps the formation of a precipitate is not an essential feature of bainite transformation. Clearly such a transformation is possible only in systems in which the diffusion rate of the solute is significantly greater than that of the solvent at the transformation temperature. Although bainitic transformations and structures have been identified in non-ferrous systems (Garwood[44]), detailed knowledge of these transformations is restricted to the iron–carbon system and related low-alloy steels.

In ferrous systems there are two major classifications of bainite structure, upper and lower bainite. The former appears to be derived from massive-martensite and the latter from acicular-martensite (Owen[45]). Initially, the distinction was made on the basis of kinetic characteristics of the transformation (Radcliffe and Rollason,[46] White and Owen[47])); upper bainite forming in low-alloy steels on isothermal transformation above about 350°C and lower bainite resulting from transformation at a lower temperature but above the M_s temperature. Later, Shackleton and Kelly[48] described the morphological and crystallographic differences between upper and lower bainite. In low-alloy steels the iron lattice transforms from f.c.c. austenite to b.c.c. ferrite by a martensitic type of transformation and the carbon diffuses either in the austenite or the ferrite to precipitate cementite, Fe_3C, or more complex carbides. Measurements of the habit plane of bainite plates in a variety of alloy steels have been made by Greninger and Troiano,[49] Smith and Mehl[50] and Bowles and Kennon.[51] The poles scatter over an extensive area of the stereographic triangle. There is a general tendency for the pole of the habit of lower-bainite (transformed below 350°C) to be in the areas occupied by the poles of the habit planes of acicular-martensite and as the transformation temperature is increased above 350°C the poles move closer to $\{111\}_P$. Thus, upper bainite probably has some structural relationship to massive-martensite and support for this view has been obtained by the study of the transition in structure which occurs when small concentrations of carbon are added to quenched iron–chromium (Ronald[52]) and iron–nickel (Goodenow and Heheman[53]). As for the data for martensite habits in alloy steels, the agreement between the measured habit planes and those predicted by the formal theories is poor even when the maximum allowable variation of the dilatation parameter, δ, is included in the calculations.

The shape change involved in the formation of upper bainite has not been measured, but all the crystallographic parameters for an iron–8% chromium–1·1% carbon alloy transformed to lower bainite have been studied in detail by Srinivasan and Wayman[54] who measured the habit plane, the specific variant of the orientation between parent and product and the magnitude of the shape strain. They found that the habit plane of the bainite plates, near $(254)_P$ which corresponds to $(\bar{1}22)_M$ in the bainitic ferrite, is different from that of martensite formed in the same alloy by quenching. The orientation relationships are also different, the close-packed planes $(111)_P$ and $(011)_M$ being about 0·5° apart and the close-packed directions being separated by about 10°. In martensite in the same alloy the close-packed directions are almost exactly parallel. The two structures differ also in the nature of the shape change. In lower bainite the surface tilt occurs only in one direction and no midrib is observed. The shear angle is about 7°, whereas in martensite in the same alloy the angle is near to 11°. Analysis of these data in terms of the formal crystallographic theories shows that, if no dilatation is assumed, the lattice invariant shear must be irrational in both plane and direction. However, if an isotropic contraction on the habit plane of 1·2% is allowed the data are consistent with the assumption that the lattice invariant shear occurs by slip on $(1\bar{1}1)_P$ and $(101)_P$ in the common $[\bar{1}01]_P$ direction, which corresponds to pencil glide on $(101)_M$ and $(112)_M$ in the $[\bar{1}\bar{1}1]_M$ direction. The data cannot be explained in terms of a lattice-invariant strain produced by any of the usual twinning modes.

The carbide precipitates in bainite occur in many different complex forms, the choice being influenced by the composition of the alloy and the temperature of transformation. Some characteristic carbide precipitates have been identified. In lower bainite Fe_3C particles form in parallel rows making an angle of 55–65° to the major axis of the ferrite plate and the orientation relationship is

$$(001)Fe_3C\|(211)_M$$

$$[100]Fe_3C\|[0\bar{1}1]_M$$

$$[010]Fe_3C\|[1\bar{1}\bar{1}]_M$$

which is the same relationship as that found between Fe_3C and the b.c.c. matrix in tempered acicular-martensite (Shackleton and Kelly[48]). The orientation relationship between austenite and bainitic ferrite in lower bainite is Kurdjumov–Sachs (Smith and Mehl[50]) and consequently the austenite, ferrite and cementite structures are uniquely related. This has been confirmed recently by the electron microscopy study of Srinivasan and Wayman.[54] The $\{112\}_M$ plane on which the Fe_3C particles precipitate on tempering martensite is also the twin plane on which the lattice invariant shear occurs during the formation of the martensite and thus it is usually assumed that the carbides precipitate at the twin interfaces. However, no evidence of twinning has been found in lower bainite and, as mentioned earlier, the crystallographic analysis of the shape change shows that twinning is not a possible lattice-invariant shear mode.

The carbide, usually Fe_3C, in upper bainite precipitates at the boundaries between the ferrite plates and, at high transformation temperatures for low-carbon alloys or at all temperatures within the range for high-carbon alloys, some coarse particles precipitate within the ferrite grains. The carbide Fe_3C precipitates in a number of different orientations to the ferrite which can be explained (Shackleton and Kelly[48]) by assuming that the Fe_3C precipitates in austenite with the Pitsch[55] orientation

$$(001)Fe_3C\|(\bar{2}25)_P$$

$$[100]Fe_3C\|[\bar{5}5\bar{4}]_P$$

$$[010]Fe_3C\|[\bar{1}\bar{1}0]_P$$

and the austenite then transforms to ferrite with one of the variants of the Kurdjumov–Sachs orientation. Smith and Mehl[50] found the austenite–ferrite orientation in upper bainite to be Nishiyama but the difference is not large enough to invalidate the Shackleton–Kelly argument.

7. GENERAL FEATURES OF THE KINETICS OF THE TRANSFORMATIONS

Bainite and martensite transformations occur at different rates. Bainite formation is relatively slow and is controlled by the rate at which the bainite plates grow, a process dependent upon the diffusion of the solute over large lattice distances. It is commonly held that, in low-alloy steels, upper bainite growth involves, but is not necessarily controlled by, the diffusion of carbon in austenite and the growth of a plate of lower bainite requires diffusion of carbon through the bainitic–ferrite lattice. The crystallographic observations appear to confirm these ideas by indicating that in upper bainite the carbide precipitates within the austenite and in lower bainite within the ferrite. However, the kinetics of plate growth in bainite is undoubtedly more complicated than suggested by these concepts. It is possible, for example, that the interaction of the carbide particles with the moving interface dislocation array may be a factor limiting the speed of bainitic growth. The kinetics of bainite formation in steels has been reviewed by Barford.[56]

Usually, the growth of martensite is extremely rapid and the overall transformation rate is determined by the frequency of nucleation of martensite plates. The nucleation event is greatly influenced by temperature so that at each temperature reached during cooling an increment of transformation occurs very rapidly but no further transformation can be induced by isothermal holding. Thus, the extent of the transformation is a unique function of the difference between the M_s and the ambient temperatures, a behaviour described as athermal. There are, however, important exceptions to this generalisation. Recently it has been shown[126] that in alloys which form massive-martensite the addition of only a very small concentration of carbon slows down the rate of growth of the plates to the point at which the growth is only a little faster than that of bainite in alloys with slightly higher concentrations of carbon. In such alloys the martensite transformation occurs isothermally with the overall kinetics determined by the growth process. The transformation in dilute alloys of uranium with transition metals (Burke[57]) is similar in that the nucleation of the α-phase occurs rapidly after an incubation period but the plates thicken slowly. In fact, the growth of the plates can be observed easily by optical microscopy at room temperature. Such isothermal growth of martensite plates is unusual, the only other authentic case which has been reported being the growth of β' martensite in copper–aluminum–nickel alloys (Hull and Garwood[58]). There is a large class of ternary and more complex alloys in which the martensitic transformation proceeds isothermally[76] but with the overall kinetics determined by the nucleation frequency. The nucleation continues over a measurable time interval although each plate, once nucleated, grows rapidly.

Since bainite formation requires long-range diffusion, the transformation can be suppressed by rapid cooling. The quenched alloy transforms martensitically at a temperature below the bainite range. Thus, by changing the cooling rate, martensite or bainite can be formed in the same alloy. However, it has not proved possible to change the structure of any alloys from acicular to massive-martensite simply by changing the cooling rate, although some change in the M_s temperature has been demonstrated at very rapid cooling rates (Swanson and Parr[59]). The change in habit and morphology of martensite occurs in many systems on changing the concentration of solute. This change is accompanied by a change in the M_s temperature. Without exception, acicular-martensite forms in the lower range of M_s temperature. The martensite transformation in the iron–nickel system is an example. The M_s temperature, which is quite insensitive to cooling rate, varies with nickel concentration (Fig. 23). The change from a habit near to $\{111\}_P$ to $\{259\}_P$ martensite occurs discontinuously at about 28·5% nickel but there is no discontinuity in the plot of M_s temperature near this composition. A similar transition from massive to acicular-martensite occurs in iron–carbon alloys, but this transition appears to be more gradual.[127] Continuity of M_s temperature on changing martensite morphology by changing solute concentration is quite general. The free energy change accompanying the martensite transformation in the iron–carbon and iron–nitrogen systems increases somewhat with decreasing M_s temperature but there is no discontinuous change in the magnitude of this energy on changing morphology (Bell and Owen[64]). The basic reason for the change in morphology is not known. The change does not occur at the same M_s temperature or at a constant degree of undercooling, or at the same value of the free-energy change in different dilute alloys of iron. It has been suggested that the variation with temperature of the critical resolved shear stresses for slip or twinning, to provide the lattice invariant strain, is an important factor. Another suggestion is that, in alloys containing carbon or nitrogen, the energy change accompanying Zener ordering can be important.[25] Recently, Nilles[135] has shown that in the iron–nickel and iron–platinum series the composition at which the

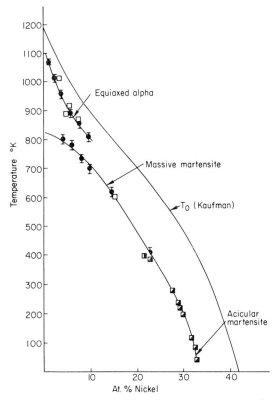

Fig. 23.   Variation of M_s temperature of iron–nickel alloys with concentration of nickel. Φ Gilbert, Wilson and Owen.[36, 60] □ Jones and Pumphrey.[61] ◢ Kaufman and Cohen.[62] ◨ Yeo.[63]

morphological change occurs coincides with the composition at which the M_s and Currie temperatures are equal. That is, acicular-martensite forms from the harder ferromagnetic austenite. Davies and Magee[136] have reported similar observations on iron–nickel and iron–nickel–cobalt alloys and they have emphasised the importance of the resistance of the austenite to plastic deformation. Resolution of this problem of the change in morphology must await the development of a more complete understanding of the nucleation and growth of martensite.

8.   GROWTH

In a single interface transformation such as occurs in gold–cadmium alloys there appears to be complete coherency between the parent and the product phases so that the interface can be moved back and forth and the transformation can be reversed (by changing from cooling to heating) with very little hysteresis. Martensitic transformation in most alloys requires substantial undercooling before spontaneous rapid growth of the martensite crystals occurs. As seen from consideration of the formal crystallographic theory, the habit planes observed to occur cannot result in an exact matching of the parent and product lattice planes at the interface. In the Bowles–Mackenzie theory these differences are accommodated by the dilatation parameter. In physical terms it is more appealing to describe the interface in terms of an array of dislocations. Then the rate of growth of a plate can be discussed in terms of the rate at which the dislocation boundary moves through the parent lattice. Bunshah and Mehl[66] have measured the time of formation of individual plates in iron–nickel–carbon alloys to be about 10^{-7} seconds, indicating that the interface, during the formation of acicular-martensite moves at about a third of the speed of sound in the solid.

Description of the interface in terms of familiar slip or twinning dislocations has proved to be difficult. One of the few models which appears to be satisfactory in many, but not all, respects was proposed by Frank[67] to describe the $\{225\}_P$ habit plane in steels. This is a degenerate case of the formal theory because the $\{225\}_P$ habit is associated with an almost exact Kurdjumov–Sachs orientation so that $[110]_P$ is parallel to $[\bar{1}1\bar{1}]_M$ and the $(111)_P$ and $(101)_M$ planes, which are parallel to each other, meet edge on at the interface. The invariant line is $[\bar{1}10]_P$. The geometrical problem is essentially two-dimensional[68] as shown in Fig. 24 in which the angles φ and ψ are defined. When $\varphi = 25 \cdot 2°$ the habit is $\{225\}_P$. The small rotation ψ is introduced to match the planes of the two lattices at the interface, because the lattice spacing differs by about 2% when $\psi = 0$. The necessary value of ψ to obtain a match varies with φ and with the c/a ratio of the tetragonal martensite lattice. However, a rotation through the angle ψ has no effect on the mismatch along the $[\bar{1}\bar{1}0]_P$ direction in the interface. One approach to this problem is to introduce the Bowles–Mackenzie dilatation parameter δ. This represents an isotropic deformation so that if δ is applied in the $[\bar{1}\bar{1}0]_P$ direction it must be applied also in the direction in the interface normal to $[\bar{1}\bar{1}0]_P$ and, consequently, ψ will be changed. In fact, to obtain a fit the lattice must now be rotated in the negative sense of ψ. Alternatively, as assumed by Frank, the rotation ψ is in the positive sense and the mismatch in the interface in a direction parallel to the dislocations is taken up by an 'assumed Poisson's ratio'. Thus, the Frank model postulates an anisotropic distortion in the interface. A detailed

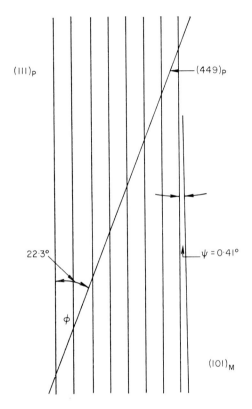

Fig. 24. (111)$_P$ and (101)$_M$ planes matching at the habit plane interface. Wayman's discussion[68] of the Frank model.[67]

critical discussion of the Frank model has been presented recently.[124,128]

For a dislocation array to be glissile it must consist of screw dislocations or there must be a component of the Burgers vector normal to the plane of the interface. In the Frank model it is assumed that when the interface moves the screw dislocations glide on {112}$_M$ planes generating the lattice invariant shear. The observed rapid motion of the interface when a martensite plate forms implies that the imposed stress is high. The limiting velocity is probably the speed of a transverse shear wave and the experimental evidence that the speed is less than the speed of sound[66] is consistent with this assumption.

Very rapid motion of the interface is possible only when the interface is separated from segregated interstitial solute impurity. Even a dilute solution of carbon in iron can have a large effect. One of the most important forms of stabilisation, depression of the M_s temperature by thermal treatment at higher temperatures, is due to the presence of carbon (Philibert[69]). This suggests that segregation of carbon to the interface of an incipient martensite plate is effective in preventing growth. When carbon is added to a pure binary iron–nickel alloy the transformation temperature decreases at first markedly with increasing carbon up to about 0·001 atom fraction (0·02 w/o) carbon but in more concentrated alloys the decrease is more gradual (Fig. 25) (Rao and Winchell[70]). A similar dramatic decrease in the transformation temperature on adding the first fraction of a percent of carbon has been observed for other binary substitutional solid-solution alloys of iron, although the concentration of carbon required to establish the variation of transformation temperature with concentration characteristic of concentrated alloys varies appreciably from one system to another. For example, in iron–chromium alloys the rate of initial decrease with increasing carbon concentration is less than for the iron–nickel series illustrated in Fig. 25 (Ronald[52]). Direct observation of the growing plates in alloys of different carbon concentration has demonstrated clearly that a small change in the concentration of carbon has a large effect on the rate of growth of the plates. A drastic retardation of growth occurs for the first small additions of carbon.[129] Over this same composition range the M_s temperature decreases markedly. Schoen[126,128] has recently developed a quantitative model to account for this inhibiting effect of carbon on the growth. The model is based on the interaction of the strain fields of carbon atoms dissolved in the martensite lattice with the stress field of the interfacial dislocation array.

When alloys are transformed at a temperature at which bainite is formed isothermally, retardation of the movement of the interface can be observed directly. When the conditions are correctly selected, the slow growth of bainite plates can be observed in a hot-stage optical microscope.[71] In principle, the slow movement of the interface could be caused by an atmosphere of interstitial atoms being dragged by the interface but no experimental support for this view is known at the present time. Instead, most recent attempts to understand the phenomenon have concentrated on the rate at which a plate lengthens and have assumed that the rate is limited by the diffusion of the interstitial away from the moving tip. A detailed quantitative model has been developed by Zener[72] and Hillert.[73] The predicted growth rates vary with concentration of solute and temperature in the sense observed experimentally, but quantitatively the agreement is not very good.[74,75]

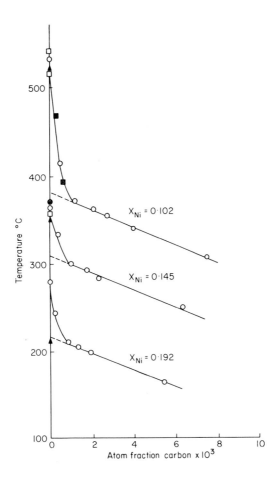

Fig. 25. Transformation start temperatures of iron–nickel–carbon alloys as a function of carbon concentration (Rao and Winchell[70]). ▲ Kaufman and Cohen.[62] □ Gilbert and Owen.[36] ● Swanson and Parr.[59] ■ Goodenow and Hehemann.[53] ○ Rao and Winchell.[70]

9. NUCLEATION

After the discovery of a martensite transformation with isothermal transformation kinetics (Kurdjumov and Maximova[76]) attempts were made to
develop a thermally activated model for the homogenous nucleation of
martensite. Considering the martensite nucleus to be a lenticular particle
lying in the habit plane and applying classical nucleation theory, Fisher,
Holloman and Turnbull[77] obtained for the saddle-point free-energy
change for the formation of a martensite nucleus

$$\Delta W^* = \frac{K\theta^2\sigma^3}{\Delta f^4} \tag{12}$$

where Δf is the chemical free-energy change per unit volume of martensite
formed, θ contains the elastic strain energy, σ is a surface energy term and
K is a numerical constant. Assuming that the energy ΔW^* is supplied by
thermal fluctuations, the nucleation frequency is

$$\dot{N} = \left(\frac{N_0}{V_M}\right) v \exp\left(-\frac{\Delta W^*}{kT}\right) \tag{13}$$

N_0/V_M is the number of atoms per unit volume and v is the lattice vib-
rational frequency. This implies that martensite forms only isothermally.
Fisher[78] considered athermal behaviour a special case of isothermal
transformation, defining M_s as the temperature at which $N = 1$ nucleus
per second. Then, using values for the variation of Δf with temperature
computed from specific heat data and a thermodynamic model for iron–
nickel alloys, it is found that

$$\frac{\Delta W^*}{kT} = 82\cdot 9 \text{ at the } M_s \text{ temperature} \tag{14}$$

and the other terms in eqn. 12 can be evaluated. In this way, Fisher calculated the isothermal nucleation rate as a function of temperature for a
series of iron–nickel alloys. The results for alloys with between 29% and
31% nickel are shown in Fig. 26. The value of N is extremely sensitive to
temperature so that the M_s temperature defined by $N = 1$ is insensitive
to cooling rate and the transformation would appear to be athermal. It is
usually supposed that isothermal nucleation frequencies can be measured
experimentally only when N is between 10^{-1} and 10^4 events per second
and, consequently, for the iron–nickel alloys considered in Fig. 26 there
is only a very narrow range of temperature and composition in which
isothermal martensite could be observed.

The inadequacies of the classical theory when applied to martensite were first demonstrated by Kaufman and Cohen.[79] Their argument can be stated by reference to the data in Fig. 26. According to Fisher's calculations no M_s temperature should be reached on cooling to 0°K in an alloy with 30·2% or more of nickel. Kaufman and Cohen showed experimentally that this is not so. In fact, an alloy with 33% nickel transforms martensitically at 4°K. It might be thought that this discrepancy could be removed by revising the values of Δf and $\theta^2\sigma^3$ but this has proved impossible and

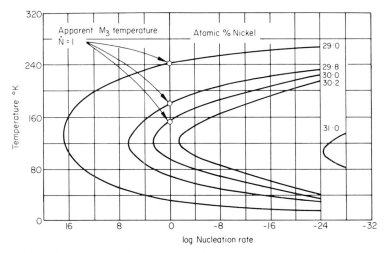

Fig. 26.   The rate of isothermal nucleation of martensite plates in iron–nickel alloys containing between 29·0 and 31·0 a/o nickel as a function of temperature, as calculated by Fisher.[78]

experience with other alloy systems has revealed even bigger discrepancies. Thus, the concept of homogeneous nucleation of martensite has been abandoned.

There exist a large number of suggestions for heterogeneous nucleation models many of which, such as the compositional fluctuation model (Fisher, Holloman and Turnbull[77]), can easily be shown to be inappropriate. However, there seems to be little doubt that the nucleation event is heterogeneous and the idea that the parent phase contains potential nuclei (embryos) with a range of potency has gained widespread support. It is usually assumed that the most potent embryos can be nucleated by a small degree of undercooling, whereas the embryos activated at the lowest temperatures are the least potent. The nucleation may or may not be assumed to be thermally activated. The concept of an embryo becoming critical at some degree of undercooling clearly allows the possibility that nucleation depends only upon temperature and is consequently athermal. There is some metallographic evidence supporting this general concept. For example, the number of plates of martensite formed on transforming a gold–47·5% cadmium alloy depends upon the temperature at which the parent phase is annealed before transformation; usually many plates form but if the annealing temperature is sufficiently high a single interface transformation occurs (Chang and Read[22]). However, attempts to find embryos in the parent crystal at a temperature above the bulk M_s temperature by electron microscopy of thin foils have been unsuccessful (Gaggero and Hull[13]).

It seems to be most reasonable to assume that the embryo is a small volume of crystal in which the lattice is strained to a configuration which approximates to the martensite lattice. These models have their origin in the reaction path theories proposed by Morris Cohen (Cohen, Machlin and Paranjpe[80]) and their culmination in the Kaufman and Cohen[81] modification of the Knapp and Dehlinger model.[82] The latter authors attempted to explain athermal nucleation in terms of the growth of an embryo with the shape of a thin oblate spheroid bounded by dislocation loops, largely of screw character, arranged to form a Frank interface. Thus, the model is applicable only to $\{225\}_P$ martensite in iron alloys. The strain energy term, 2×10^{10} ergs cm^{-3} of martensite, is numerically similar to the same term discussed by Fisher et al.[77] for homogeneous nucleation and the interfacial energy, estimated from the dislocation model, is 200 erg cm^{-2} which is an order of magnitude larger than the earlier estimates. If

$$\Delta W = \Delta f + \Delta g \tag{15}$$

where Δg is the non-chemical free-energy accompanying the formation of unit volume of martensite, the chemical free-energy difference Δf is always negative below T_0 (the temperature at which the chemical free-energy of the parent and martensitic phases are equal) and Δg is positive at all temperatures. Below T_0, ΔW first increases and then decreases rapidly with increasing size of the embryo. Knapp and Dehlinger[82] abandoned the classical assumption that nucleation occurs when the overall free-energy starts to decrease with increasing embryo size and assumed that nucleation always occurs from existing large embryos, the critical event being the cataclysmic growth which was assumed to occur when $\Delta W = 0$. This criterion appears to be purely intuitive. The size of embryo participating in the critical event is even larger than the critical size predicted by the classical theory. Using the data for iron–nickel already quoted, the nucleus at the M_s temperature contains about 8×10^6 atoms. Smaller embryos will not transform because of the large value of Δg, but these become active at some lower temperature at which Δf is sufficiently large. Thus, the athermal kinetics is explained in terms of the distribution of the size of the embryos.

Kaufman and Cohen[81] extended the model to account for isothermal nucleation. They reformulated the change of total free-energy with embryo size in terms of the expansion of a hypothetical 'giant' dislocation loop, considered to be equivalent to the array of screw dislocations in a Knapp and Dehlinger embryo larger than the classical critical size. Physically, the critical event is thought to be the formation of a new dislocation loop in the parent phase at the tip of the embryo. The radius of the equivalent giant loop at the critical event is $r_c = (36/25)r_0$, where r_0 is the size at which the embryo grows cataclysmically in the Knapp and Dehlinger model. It is assumed that if the size is greater than that at which the total free-energy decreases with increasing size (the classical radius), it can be thermally activated to the critical size. The quantitative predictions of this model were compared with the isothermal nucleation frequency data of Shih, Averbach and Cohen.[83] Good agreement was obtained after a small adjustment of the value for the surface energy.

Shih measured nucleation frequencies at the start of transformation before any athermal martensite had formed and before rapid nucleation due to the autocatalytic 'burst' phenomena had started. The values of the theoretical parameters obtained by fitting Shih's data can be used in the reverse calculation to obtain the number and distribution of embryo size in the parent phase at temperatures above T_0. The results of such calculations indicate that many embryos large enough to be seen by electron microscopy should exist and the failure to observe them constitutes one of the major objections to the theory. It has been reported (Pati and Cohen[84]) that new experimental measurements of a more sophisticated kind than these reported by Shih indicate the conclusion that, while the embryos may be large enough to view in the electron microscope, the probability of finding an embryo in any field of observation is vanishingly small due to the very small concentration of embryos.

The Kaufman and Cohen model drew attention to the possibility that the critical event in nucleation may be rapid movement of the interface of an embryo which, in the classical sense, is already supercritical. There have been several recent suggestions that the kinetics may be controlled by the rate of movement of the interface throughout the nucleation and the growth of the plate. It has been proposed that the resistance to the motion of the interface is provided by the resistance to deformation of the parent phase (Magee[85] and Breinan[86]). It is supposed that this resistance is reflected directly by a change in the strength of the martensite. For example, it is thought that Δf at M_s increases with increasing nickel content in iron–nickel alloys and that this is a direct result of increased resistance to motion of the interface due to the increased strength of the martensite with higher nickel content and lower M_s temperature. Appreciable plastic deformation of the parent phase is expected in many martensitic transformations. The resulting incremental increase in strength of the parent is inherited by the martensite. Perhaps the strength of the martensite itself is unimportant if either the lattice invariant shear or the accommodation deformation occurs after the product crystal structure is formed.[130]

The idea that the critical nucleating event is the growth of a large embryo with a structure closely related to that of the final martensite plate is a useful concept but its validity has not yet been established beyond question. However, there is little doubt that the nucleation is heterogeneous and that the nucleation sites are some form of lattice defect in the parent phase. Zener[89] suggested that the first step in the f.c.c. to b.c.c. transformation is an $a/12 \langle 112 \rangle$ shear on the $\{111\}_P$ planes followed by some reshuffling of the atoms. Later, Jaswon[90] proposed a similar model involving the dissociation of a Shockley partial into two equal 'quarter-dislocations' separated by a ribbon of fault with a structure resembling the b.c.c. lattice. Further elaboration of this idea by Bogers and Burgers[91] suggested a sequence of dissociations which could produce the complete lattice transformation. They proposed that a $a/12 \langle 112 \rangle$ shear occurs on $\{111\}_P$ followed by a shear on $\{110\}_M$ which converts the lattice to b.c.c. Electron microscopy studies of the transformation in iron–32·3% nickel by Dash and Brown[92] and in iron–33·1% nickel by Magee[131] have produced convincing evidence that a dissociation on $\{111\}_P$ is indeed involved in the nucleation. Figure 27 is taken from the work of Dash and Brown. It shows a faulted layer on $\{111\}_P$ connected to the tip of a martensite plate. A fundamental difficulty with this, as with many other similar electron microscope studies, is that there is no way of deciding whether the formation of the $\{111\}_P$ fault precedes or follows the formation of the martensite plate. It is clear, however, that a shear on $\{111\}_P$ is associated with the formation of the plate. Dash and Brown showed that the volume free-energy change, Δf, accompanying the transformation is sufficient to provide a shear stress comparable with the yield stress of the parent crystal but, unfortunately, they were not able to determine the Burgers vector of the partials associated with the $\{111\}_P$ fault.

Electron microscope studies of martensite in 18–8 stainless steel have revealed a similar situation with the addition that, in this case, a transition

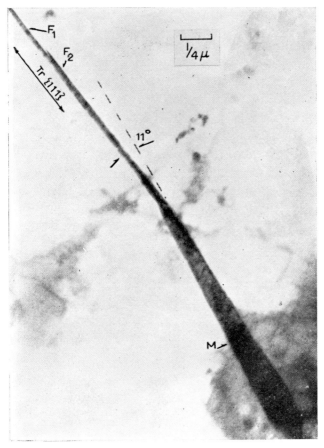

Fig. 27. Electron transmission micrograph of an iron–32·3% nickel alloy transformed at −39°C (Dash and Brown[92]).

structure has been identified. The ε-phase forms, probably athermally, by a shear on $\{111\}_P$ to give an orientation $\{111\}_P \| (0001)_\varepsilon$ by a dissociation to separated Shockley partials in a manner similar to that first described by Christian[93] for the transformation in cobalt. The plates of ε-phase observed by electron microscopy may be as thick as $0·1\mu$ and consequently there must be some mechanism operating which repeats the single layer fault over many adjacent $\{111\}_P$ planes. There is substantial evidence that α-martensite is formed by mutual interaction of plates of the ε-phase (Venables,[94] Cina,[95] Reed,[96] Breedis and Robertson[97]). Venables showed that the α-martensite first forms as needles along the line of intersection ($\langle 1\bar{1}0 \rangle_P$) of two plates of ε and that these subsequently develop into plates with a $\{225\}_P$ habit. The sequence described by Venables appears to offer a satisfactory description of the nucleation events in stainless steel, but there are some difficulties with this model. As in iron–nickel, it is not possible to be sure that the formation of stacking faults and ε-phase precedes the formation of martensite. In addition, there is some experimental evidence that in some cases ε-phase and α-martensite can form independently (Dash and Otte[98]).

It is noticeable that the plane lattice faults which have been considered as possible heterogeneous-nucleation sites do not occur on the same plane as the observed habit plane of the martensite. This is illustrated by Fig. 27. Thus, at best, consideration of the faults can provide only a partial explanation of the nucleation events. A possible exception is the nucleation of massive-martensite with a habit close to $\{111\}_P$. A second difficulty is that the critical event, presumably the dissociation, does not coincide with the critical situation required by classical theories of heterogeneous nucleation. Further, the dislocation dissociation models suggest that nucleation of martensite should be easier in alloys with low stacking-fault energy. To date, there is no clear evidence that this is so.

At the most fundamental level, the nucleation of martensite is a problem in relative lattice stability (Zener[89]). A useful way of approaching these problems is by consideration of the elastic constants. The lattice is considered to become unstable when the elastic anisotropy, $2C_{44}/(C_{11} - C_{12})$, reaches some high value. A quantitative model in these terms has not yet been devised. This is due in part to the sparsity of accurate knowledge of the variation with temperature and pressure of the C_{44} and $(C_{11} - C_{12})/2$ elastic constants of the parent and product phases. Relevant data are beginning to appear (Fisher[99,100]). Robertson[101] has pointed out that the undercooling $(T_0 - M_s)$ in iron–nickel alloys increases as the moduli of the parent phase increase. He attributed this to the increase in restraint imposed on the transformation by the surrounding matrix but no quantitative treatment was developed. Recently, however, it has been shown[132] that for iron–30 nickel the anisotropy factor changes by only 2% in the 100°C temperature interval above the M_s temperature. Thus, it must be concluded that Zener's high anisotropy hypothesis represents a necessary but no sufficient condition for lattice instability and transformation.

10. THE PROPERTIES OF MARTENSITIC MATERIALS

Although all the structure-sensitive physical properties change when a material undergoes a martensitic transformation, the only change of property which has been studied in detail is the change in strength (hardness or flow stress) in iron alloys. Again, in the modern work emphasis has been placed on studies of iron–nickel–carbon alloys and a few of the results obtained will be discussed here. There has been some work on the fracture characteristics of martensitic steels but these investigations have been concerned with developing engineering knowledge of the properties of commercial steels and little work on the fundamentals of cleavage in simple martensitic alloys has appeared. A start has been made on the study of the effect of a martensitic transformation on other properties. In particular, there is interest in the transformation in hard superconducting alloys such as V_3Si because the crystallographic transformation may be intimately related to the transition in superconducting properties. However, the crystallographic transformations are not sufficiently characterised to warrant a discussion of this phenomenon at this time.

There are two basic reasons why the flow stress of iron–nickel–carbon alloys can be increased by a martensitic transformation. One is the almost incidental fact that to keep a high concentration of carbon in solution it is necessary to quench rapidly from a temperature at which the austenite (f.c.c.) phase is stable and, in many alloys, a martensitic transformation occurs during the quench. The second reason is that the martensitic transformation produces a characteristic substructure, either a high density of dislocations or fine parallel twins, and the hardening mechanisms involve interactions with this substructure.

It is customary to list the various features of the structure and the strengthening mechanisms which contribute to the strength of an alloy transformed to martensite, although it should not be assumed that these are additive. Such lists include hardening by refinement of the size of the structural unit (an effect related to the size of the martensite plates or the massive blocks), substructure, substitutional solid-solution, interstitial solution-hardening and precipitation hardening. Modern interest in these effects dates from a series of papers by Winchell and Cohen starting in 1962.[102–105] They attempted to isolate individual hardening effects and to study them quantitatively. They showed how to obtain a meaningful value of the flow stress of the martensite in a specimen containing a significant volume-fraction of retained austenite and they emphasised the importance of studying 'virgin' martensite, that is martensite in which no precipitation has occurred either during the quench or subsequently. Probably the most important results of this work were the recognition of the large contribution to the strength made by interstitial solute-hardening in alloys containing an appreciable concentration of carbon and the importance of the substructure in the solution-hardening mechanism. They considered only the flexibility of the moving dislocation line as modified by the presence of a twinned substructure but the basic experimental approach has been extended by others to include the study of martensites with a high density of dislocations. [107]

The direct effect of substructure on the flow stress of martensites can be studied in iron–nickel alloys provided the solid-solution hardening due to the nickel is taken into account. The flow stress increases rapidly with the addition of the first 4% of nickel but in more concentrated alloys there is only a very small increase with increasing nickel concentration up to 33 a/o. This behaviour appears to follow the change in lattice parameter with increasing nickel content. No discontinuity in the shear modulus variation with nickel concentration occurs over the range 0% to 33% nickel (Roberts and Owen[106]). The substructure of martensite as observed by electron microscopy does not change in any detectable way between 4% and 28·5% and thus it can be concluded that hardening due to substructure is constant over this range and the magnitude of this hardening can be deduced by comparing the flow stress of martensite with that of fully annealed iron–nickel ferrite (Speich and Swann[38]). About 60% of the flow stress of a nearly pure iron in the martensitic state can be attributed to the substructure. This increment of stress does not change appreciably on adding nickel up to 33% nickel. In iron–nickel alloys with more than 4% nickel, about 40% of the hardening is due to the substructure and 35% is due to solution hardening by nickel. The contribution of substructure hardening to the total strength is much less significant in iron–nickel–carbon martensite because interstitial solution and precipitation hardening predominate. The change from a dislocation to a twin substructure in the binary iron–nickel alloys occurs at about 28·5% nickel. The flow stress does not change significantly on increasing the concentration of nickel through this concentration and consequently it appears that the strengthening due to a dislocation substructure is about the same as that due to fine stacks of twins.

As first pointed out by Winchell and Cohen[103] for twinned structures and Roberts and Owen[107] for martensite with a dislocation substructure, the flow stress at, say 0·6% total strain and a constant temperature is a linear function of the square root of the atom fraction of carbon in solution. The slope is nearly but not quite the same for the two substructures (Fig. 28) and is that predicted by all the models which assume that the hardening

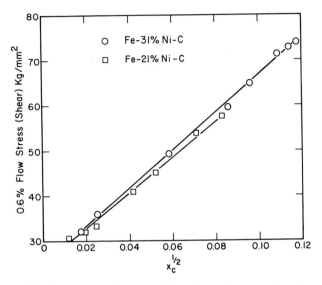

Fig. 28.   0·6% flow stress of iron–nickel–carbon alloys as a function of square root of the atom fraction of carbon (Roberts and Owen[109]).

is a result of the elastic strain interaction between the dislocation and the lattice distortion produced by an interstitial carbon atom. Fleischer and Hibbard[108] pointed out that the interaction is particularly strong when the solute atom produces a tetragonal distortion and they developed a quantitative model for this type of hardening. This theory predicts the slopes of the plots in Fig. 28 remarkably well (Roberts and Owen[109]), but the theory is not applicable to solution hardening by carbon in martensite because, in common with all other theories involving short-range elastic interaction between individual carbon atoms and the glide dislocations, it requires that the flow stress decrease with increasing testing temperature. This is a direct result of an intrinsic assumption of the theory that the interaction is thermally activated. An increase in flow stress with decreasing temperature is, in fact, observed but the thermal effect is not associated with the presence of the carbon atoms. This is demonstrated by the fact that the slopes of the plots in Fig. 28 are unchanged by changing the testing temperature. The point has been confirmed by extensive measurements of the activation parameters for plastic flow as a function of temperature and carbon concentration.[109] Thus, carbon hardens iron–nickel–carbon martensite by increasing the long-range stress field, but no satisfactory model has been developed. The decrease of the flow stress with increasing temperature is probably due to thermal activation of the glide dislocations over the Peierls barrier by a double-kink mechanism (Roberts and Owen[109]).

At temperatures below $-50°C$ the variation of the activation enthalpy with temperature is linear, indicating that plastic flow is controlled by a single mechanism. At higher temperatures this enthalpy increases steeply but non-linearly with temperature indicating a more complex situation. Diffusion of carbon atoms involving one or more jumps can take place in a few seconds at temperatures above $-50°C$ and segregation of the carbon atoms can occur during plastic deformation (Winchell and Cohen[103]). At temperatures between $40°C$ and $20°C$ evidence of Snoek interaction has been found (Owen and Roberts[110]) and at higher temperatures Cottrell atmospheres can form around stationary dislocations or, at the appropriate temperature and strain rate, the glide dislocations can drag Cottrell atmospheres through the lattice (Owen and Roberts[111]). Atmospheres moving with the glide dislocations cause serrated flow and a negative strain-rate sensitivity of the flow stress.

Recently, there has been much interest in a phenomenon which has come to be known as the S–D effect. In a wide range of steels, martensite is much stronger in uniaxial compression than in uniaxial tension.[137] A number of explanations have been proposed to account for this surprising phenomenon.[138] Among the most promising are the microcrack, the residual stress, the Bauschinger and the solute-dislocation interaction hypotheses. A special case of the latter hypothesis involving a model in which the elastic strains are reasonably considered to be non-linear has been discussed in some detail by Hirth and Cohen.[138] In our opinion, the classical explanation that the strength differential is due to the formation of microcracks, either during the quench or during the testing in uniaxial tension, is the most widely applicable model.

An additional increment of strength is obtained when fine precipitates form. The necessary conditions are determined by alloy composition, temperature and time. In martensitic steels deformed at temperatures above about $100°C$ complex interactions can occur between the glide dislocations, the atmospheres, the solute in the lattice and precipitate particles (Kalish and Cohen[112]).

11. ACKNOWLEDGEMENTS

We are grateful to Dr G. R. Srinivasan, and Dr J. Nilles for permission to use some of their results and photomicrographs before publication. Part of the work described here was supported by the Advanced Research Projects Agency through the Materials Science Center at Cornell University.

12. REFERENCES

1. Frank, F. C. (1963). *NPL Conference on the Relation Between Structure and Strength in Metals and Alloys*, HMSO, London, p. 248.
2. Reed, R. P. and Breedis, J. F. (1966). In *Behavior of Metals at Cryogenic Temperatures*, ASTM STP 387, Am. Soc. Testing Mats., p. 60.
3. Christian, J. W. (1965). *Theory of Transformation in Metals and Alloys*, Chap. XXII, Pergamon, London.
4. Wayman, C. M. (1964). *Introduction to the Crystallography of Martensitic Transformations*, Macmillan, New York.
5. Greninger, A. B. and Troiano, A. R. (1949). *Trans. AIME*, **185**, 590.
6. Brook, R. (1965). *Physical Properties of Martensite and Bainite, Iron and Steel Institute Special Report* 93, p. 165.
7. Bain, E. C. (1924). *Trans. AIME*, **70**, 25.
8. Jaswon, M. A. and Wheeler, J. A. (1949). *Acta Cryst.*, **1**, 216.
9. Bell, T. and Owen, W. S. (1967). *J. Iron and Steel Inst.*, **205**, 428.
10. Crocker, A. G. and Bilby, B. A. (1961). *Acta Met.*, **9**, 678.
11. Warlimont, H. (1965). *Physical Properties of Martensite and Bainite, Iron and Steel Institute Special Report* 93, p. 58.
12. Bowles, J. S. and Mackenzie, J. K. (1954). *Acta Met.*, **2**, 138.
13. Gaggero, J. and Hull, D. (1962). *Acta Met.*, **10**, 995.
14. Lieberman, D. S. (1958). *Acta Met.*, **6**, 680.
15. Christian, J. W. (1955–56). *J. Inst. Metals*, **84**, 349.
16. Wechsler, M. S. (1959). *Acta Met.*, **7**, 793.
17. Kurdjumov, G. and Sachs, G. (1930). *Z. Physik.*, **64**, 325.
18. Nishiyama, Z. (1934–35). *Sci. Rep. Tohoku Univ.*, **23**, 638.
19. Wechsler, M. S., Lieberman, D. S. and Read, T. A. (1953). *Trans. AIME*, **197**, 1503.
20. Mackenzie, J. K. and Bowles, J. S. (1957). *Acta Met.*, **5**, 137.
21. Lieberman, D. S., Wechsler, M. S. and Read, T. A. (1955). *J. App. Phys.*, **26**, 473.
22. Chang, L. C. and Read, T. A. (1951). *Trans. AIME*, **191**, 516.
23. Burkhart, M. W. and Read, T. A. (1953). *Trans. AIME*, **197**, 516.
24. Efsic, E. J. and Wayman, C. M. (1967). *Trans. Met. Soc., AIME*, **239**, 873.
25. Owen, W. S., Wilson, E. A. and Bell, T. (1965). In *High Strength Materials*, Ed. by V. Zackay, Wiley, New York, p. 167.
26. Kelly, P. M. and Nutting, J. (1960). *Proc. Roy. Soc.*, **A259**, 45; *J. Iron and Steel Inst.* (1961), **197**, 199.

27. Patterson, R. L. and Wayman, C. M. (1964). *Acta Met.*, **12**, 1306.
28. Otte, H. M. and Read, T. A. (1957). *Trans. AIME*, **209**, 412.
29. Johnson, K. A. and Wayman, C. M. (1963). *Acta Cryst.*, **16**, 480.
30. Otte, H. M. (1960). *Acta Met.*, **8**, 892.
31. Crocker, A. G. and Bilby, B. A. (1961). *Acta Met.*, **9**, 678 and 992.
32. Sauver, A. and Chon, C. H. (1929). *Trans. AIME*, **84**, 350.
33. Mehl, R. F. and Smith, D. W. (1934). *Trans. AIME*, **113**, 203.
34. Mehl, R. F. and Derge, G. (1937). *Trans. AIME*, **125**, 482.
35. Marder, A. R. and Krauss, G. (1967). *Trans. ASM*, **60**, 651.
36. Gilbert, A. and Owen, W. S. (1962). *Acta Met.*, **10**, 45.
37. Srinivasan, G. R. and Owen, W. S., to be published.
38. Speich, G. R. and Swann, P. R. (1965). *J. Iron and Steel Inst.*, **203**, 480.
39. Warlimont, H. (1961). *Trans. Met. Soc. AIME*, **221**, 1270.
40. Hull, D. (1962). *Phil. Mag.*, **7**, 537.
41. Nishiyama, Z. and Shimizu, K. (1959). *Acta Met.*, **7**, 432.
42. Srinivasan, G. R. (1967). Discussion in *Transformation and Hardenability in Steel, Climax Molybdenum*, Ann Arbor, p. 192.
43. Klostermann, J. A. (1965). *Physical Properties of Martensite and Bainite, Iron and Steel Institute Special Report* 93, p. 20.
44. Garwood, R. D. (1965). *Physical Properties of Martensite and Bainite, Iron and Steel Institute Special Report* 93, p. 90.
45. Owen, W. S. (1961). *J. Iron and Steel Inst.*, **198**, 170.
46. Radcliffe, S. V. and Rollason, E. C. (1959). *J. Iron and Steel Inst.*, **191**, 56.
47. White, J. S. and Owen, W. S. (1960). *J. Iron and Steel Inst.*, **195**, 79.
48. Shackleton, D. S. and Kelly, P. M. (1965). *Physical Properties of Martensite and Bainite, Iron and Steel Institute Special Report* 93, p. 126.
49. Greninger, A. B. and Troiano, A. R. (1940). *Trans. AIME*, **140**, 307.
50. Smith, G. V. and Mehl, R. F. (1942). *Trans. AIME*, **150**, 211.
51. Bowles, J. S. and Kennon, N. A. (1960). *J. Aust. Inst. Met.*, **5**, 106.
52. Ronald, T. M. F. (1964). Ph.D. Thesis, University of Liverpool.
53. Goodenow, R. H. and Heheman, R. F. (1965). *Trans. AIME*, **223**, 1777.
54. Srinivasan, G. R. and Wayman, C. M. (1968). *Acta Met.*, **16**, 609, 621.
55. Pitsch, W. (1962). *Acta Met.*, **10**, 897 and *Arch. Eisenh*, (1963), **34**, 381.
56. Barford, J. (1966). *J. Iron and Steel Inst.*, **204**, 609.
57. Burke, J. (1965). *Physical Properties of Martensite and Bainite, Iron and Steel Institute Special Report* 93, p. 83.
58. Hull, D. and Garwood, R. D., Institute of Metals Monograph and Report Series, 18 (1955) p. 219 and *J. Inst. Met.*, (1957/58), **86**, 485.
59. Swanson, W. D. and Gordon Parr, J. (1964). *J. Iron and Steel Inst.*, **202**, 104.
60. Wilson, W. A. (1964). Ph.D. Thesis, University of Liverpool.
61. Jones, F. W. and Pumphrey, W. I. (1949). *J. Iron and Steel Inst.*, **163**, 121.
62. Kaufman, L. and Cohen, Morris (1956). *Trans. AIME*, **206**, 1393.
63. Yeo, R. G. (1962). *Trans. AIME*, **224**, 1222; *Trans. ASM* (1964), **57**, 48.
64. Bell, T. and Owen, W. S. (1967). *Trans. AIME*, **239**, 1940.
65. Lieberman, D. S. (1955). *The Mechanism of Phase Transformation in Metals, Institute of Metals Monograph and Report*, 18.
66. Bunshah, R. F. and Mehl, R. F. (1953). *Trans. AIME*, **197**, 1251.
67. Frank, F. C. (1953). *Acta Met.*, **1**, 15.
68. Wayman, C. M. (1961). *Acta Met.*, **9**, 912.
69. Philibert, J. (1955). *Compt. Rend.*, **240**, 190.
70. Rao, M. M. and Winchell, P. G. (1967). *Trans. AIME*, **239**, 956.
71. Speich, G. R. (1961). In *Decomposition of Austenite by Diffusional Processes*, Interscience, New York, p. 353.

72. Zener, C. (1946). *Trans. AIME*, **167**, 550.
73. Hillert, Mats (1957). *Jernkontorets Ann.*, **141**, 757.
74. Speich, G. R. and Cohen, Morris (1960). *Trans. AIME*, **218**, 1050.
75. Kaufman, L., Radcliffe, S. V. and Cohen, Morris (1961). In *Decomposition of Austenite by Diffusional Processes*, Interscience, New York, p. 313.
76. Kurdjumov, G. V. and Maximova, P. P. (1948). *Dokl. Akad. Nank USSR*, **61**, 83.
77. Fisher, J. C., Holloman, J. H. and Turnbull, D. (1947). *Trans. AIME*, **185**, 691.
78. Fisher, J. C. (1953). *Trans. AIME*, **197**, 918.
79. Kaufman, L. and Cohen, Morris (1956). *The Mechanism of Phase Transformation in Metals, Institute of Metals Monograph and Report Series*, 18, p. 18.
80. Cohen, Morris, Machlin, E. S. and Paranjpe, V. G. (1949). *Thermodynamics in Physical Metallurgy*, ASM, p. 242.
81. Kaufman, L. and Cohen, Morris (1958). *Prog. Metal Phys.*, **7**, 165.
82. Knapp, H. and Dehlinger, U. (1956). *Acta Met.*, **4**, 289.
83. Shih, C. H., Averbach, B. L. and Cohen, Morris (1955). *Trans. AIME*, **203**, 183.
84. Pati, S. R. and Cohen, Morris (1969). *Acta Met.*, **17**, 189.
85. Magee, C. L. (1966). Ph.D. Thesis, Carnegie Institute of Technology.
86. Breinan, E. M. (1968). Ph.D. Thesis, Rensselaer Polytechnic Institute.
87. Yeo, R. B. G. (1964). *Trans. ASM*, **57**, 48.
89. Zener, C. (1948). *Elasticity and Anelasticity of Metals*, Chicago University Press.
90. Jaswon, M. A. (1956). *The Mechanism of Phase Transformation in Metals, Institute of Metals Monograph and Report Series*, 18, p. 173.
91. Bogers, A. J. and Burgers, W. G. (1964). *Acta Met.*, **12**, 255.
92. Dash, S. and Brown, N. (1966). *Acta Met.*, **14**, 595.
93. Christian, J. W. (1951). *Proc. Roy. Soc.*, **A206**, 51.
94. Venables, J. A. (1962). *Phil. Mag.*, **7**, 35.
95. Cina, B. (1954). *J. Iron and Steel Inst.*, **177**, 406.
96. Reed, R. P. (1962). *Acta Met.*, **10**, 865.
97. Breedis, J. F. and Robertson, W. D. (1962). *Acta Met.*, **10**, 1077.
98. Dash, J. and Otte, H. M. (1962). *Proc. Fifth International Congress on Electron Microscopy*, Academic Press, New York.
99. Fisher, E. S. and Renken, C. J. (1964). *Phys. Rev.*, **A482**, 135.
100. Fisher, E. S. and Dever, D. (1967). *Trans. Met. Soc. AIME*, **48**, 239.
101. Robertson, W. D. (1965). *Physical Properties of Martensite and Bainite, Iron and Steel Institute Special Report* 93, p. 26.
102. Winchell, P. G. and Cohen, Morris (1962). *Trans. ASM*, **55**, 347.
103. Winchell, P. G. and Cohen, Morris (1963). In *Electron Microscopy and the Strength of Crystals*, Interscience, New York.
104. Cohen, Morris (1962). *Trans. AIME*, **224**, 638.
105. Cohen, Morris (1963). *J. Iron and Steel Inst.*, **201**, 833.
106. Roberts, M. J. and Owen, W. S. (1967). *Trans. ASM*, **60**, 687.
107. Roberts, M. J. and Owen, W. S. (1965). *Physical Properties of Martensite and Bainite, Iron and Steel Institute Special Report* 93, p. 171.
108. Fleischer, R. L. and Hibbard, W. R. (1963). *Conference on the Relation between Structure and the Mechanical Properties of Metals, NPL Symposium*, HMSO, London, p. 262.
109. Roberts, M. J. and Owen, W. S. (1968). *J. Iron and Steel Inst.*, **206**, 375.
110. Owen, W. S. and Roberts, M. J. (1948). In *Dislocation Dynamics*, Wiley, New York.
111. Owen, W. S. and Roberts, M. J. (1967). *Proceedings of International Conference on Strength of Metals and Alloys*, Tokyo, Japan.
112. Kalish, D. and Cohen, Morris (1968). To be published.
113. Machlin, E. S. and Cohen, Morris (1951). *J. Metals*, **3**, 1019.

114. Entwistle, A. R. (1956). *The Mechanism of Phase Transformations In Metals,* Inst. of Metals, p. 315.
115. Bowles, J. S. (1951). *Acta Cryst.,* **4,** 162.
116. Lomer, W. M. (1956). *The Mechanism of Phase Transformations in Metals,* Institute of Metals, p. 337.
117. Krauklis, P. and Bowles, J. S. (1969). *Acta Met.,* **17,** 997.
118. Oka, M. and Wayman, C. M. (1969). *Trans. ASM,* **62,** 370.
119. Rowlands, P. C., Fearon, E. O. and Bevis, M. (1969). *The Mechanism of Phase Transformations in Crystalline Solids,* Inst. Met. Monograph 33, London, p. 164.
120. Acton, A. F. and Bevis, M. (1969). *Mat. Scien. and Eng.,* **5,** 19.
121. Ross, N. D. H. and Crocker, A. G. (1970). *Acta Met.,* **18,** 405.
122. Dunne, D. P. and Bowles, J. S. (1969). *Acta Met.,* **17,** 201.
123. Wechsler, M. S., Read, T. A. and Lieberman, D. S. (1960). *Trans. AIME,* **218,** 202.
124. Owen, W. S., Schoen, F. J. and Srinivasan, G. R. (1970). *Phase Transformations,* ASM Seminar Proceedings, 157.
125. Schoen, F. J., Nilles, J. L. and Owen, W. S. To be published.
126. Schoen, F. J. and Owen, W. S. To be published in *Met. Trans.*
127. Wayman, C. M. (1968). *Adv. Mat. Res.,* **3,** 147.
128. Schoen, F. J. (1970). Ph.D. Thesis, Cornell University.
129. Goodenow, R. H., Matas, S. J. and Hehemann, R. F. (1963). *Trans. AIME,* **227,** 651.
130. Pascover, S. J. and Radcliffe, S. V. (1969). *Acta Met.,* **17,** 321.
131. Magee, C. L. (1970). *Phase Transformations,* ASM Seminar Proceedings, 115.
132. Salama, K. and Alers, G. A. (1968). *J. Appl. Phys.,* **39,** 4857.
133. Ko, T. and Cottrell, S. A. (1952). *J. Iron and Steel Inst.,* **172,** 307.
134. Ko, T. (1953). *J. Iron and Steel Inst.,* **173,** 16.
135. Nilles, J. (1970). Ph.D. Thesis, Cornell University.
136. Davies, R. G. and Magee, C. L. (1970). To be published in *Met. Trans.*
137. Leslie, W. C. and Sober, R. J. (1967). *ASM Trans. Quart.,* **6,** 459.
138. Hirth, J. P. and Cohen, Morris (1970). *Met. Trans.,* **1,** 3.

CHAPTER 5

DEVELOPMENT OF MICROSTRUCTURE

R. B. NICHOLSON AND G. J. DAVIES

1. INTRODUCTION

It is through microstructure that the physical world of interatomic forces
is linked to the reality of the mechanical properties of materials. In principle,
it is possible to calculate the mechanical properties of single crystals of
pure substances and simple compounds from a knowledge of their crystal
structure and the nature of their interatomic forces; in practice, this can
usually only be done for the elastic constants of some pure substances.
For engineering materials, however, even if such calculations were possible they would be of little value since the major variable which dictates
the range of properties obtainable is *microstructure*. Microstructure is
important both in the form of the arrangement of grains in a single-phase
polycrystal and in the arrangement of phases in a multiphase material.
Properties which depend on microstructure are said to be *structure-
sensitive* and include the electrical, magnetic and plastic properties of
materials. Examples of properties which are generally insensitive to
structure are the elastic modulus, the specific heat and the corrosion
resistance, although the first is certainly structure-sensitive in fibre-
composites and textured materials, and the last can depend on structure
in some forms of corrosion, *e.g.* stress corrosion and intergranular
corrosion.

Microstructure is, therefore, as important in determining properties as
are the atoms from which the material is constructed. In examining
microstructure we are normally dealing with aggregates of many millions
of atoms although both smaller and larger aggregates must sometimes be
considered. Hence we are concerned with dimensions from about 0·01 μ
to 100 μ. Dimensions greater than 100 μ are sometimes important and
features on this scale can be referred to as *macrostructural*.

Studies of microstructure have been carried out by scientists interested in materials for well over a hundred years. These studies have been made using, in the main, the light microscope, with reflected light for metals and with both reflected and transmitted light for minerals. More recently, X-ray diffraction, electron microscopy and field-ion microscopy have become widely used as the aggregates have been studied at higher and higher magnifications.

Microstructure results initially from the phase change which causes a liquid to crystallise as the temperature is reduced, and subsequently, from any solid state changes which may occur. Microstructural features can also be generated by the plastic deformation of some materials, particularly metals. All the changes are controlled by the same principles of thermodynamics and kinetics. Consequently we begin this review with a general discussion of the thermodynamics of microstructure. Microstructural development by solidification and by changes in the solid state are then considered separately. Under both headings some consideration has been given to the development of macrostructural features.

2.   THE THERMODYNAMICS OF MICROSTRUCTURE

Before we can describe the details of the transformations involved in the development of microstructure we must consider the reasons for transformation. In general, we are concerned with a change from one position of stable or metastable equilibrium to another in response to a driving force. The existence of a driving force, *viz.* a decrease in the free energy of the system, indicates that a transformation is favourable. This driving force is a necessary but not sufficient requirement. Whether or not a transformation takes place is determined by kinetic factors. First, we must ask can the transformation begin? To answer this we turn our attention to the study of nucleation. Secondly, we must ascertain that after nucleation the transformation can continue. This involves the study of growth. Of course, there are particular transformations, *e.g.* spinodal decomposition (*see* Section 2.5) which do not involve this classical sequence of nucleation and growth. In these cases the nucleation barrier is absent and growth phenomena are of primary importance.

In this section we shall first consider the nature of the driving force for a transformation and then consider kinetic effects. This will involve a study of the theories of nucleation and an outline of the differences between those transformations which depend on a nucleation phenomenon and those which do not. We will consider the kinetics of growth subsequently in Sections 3 and 4.

2.1. Free energy changes

When considering the development of microstructure we are concerned with heterogeneous equilibrium, *i.e.* equilibrium involving more than one phase. The free energy, G, of a component phase is defined by

$$G = H - TS \tag{1}$$

where H is the enthalpy, T the absolute temperature and S, the entropy. For most metallurgical systems pressure can be considered to be constant so that

$$\left(\frac{\partial G}{\partial T}\right)_{p=\text{constant}} = -S \tag{2}$$

Thus the free energy decreases with increasing temperature.

The variation of the free energies of the different pure metal phases is shown schematically in Fig. 1. The change in free energy on transformation at constant temperature from one phase to another is given by

$$\Delta G = \Delta H - T\,\Delta S \tag{3}$$

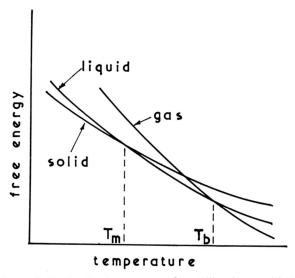

Fig. 1.   The variation in the free energy of metallic phases with temperature (schematic).

At equilibrium between two phases,

$$\Delta G = 0$$

and this defines the equilibrium melting point, T_m, and boiling point, T_b. At other temperatures, the equilibrium phase is that which has minimum free energy. ΔG, the difference between the free energies provides the driving force for transformation.

With simple binary alloys the free energy can vary with composition in several ways, as shown in Fig. 2. The free energy curves for the component phases move relatively as the temperature changes. For a particular

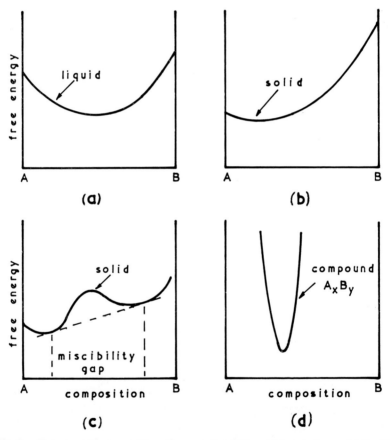

Fig. 2.   Free-energy composition diagrams for different binary alloys: (a) liquid, (b) solid, (c) solid with a miscibility gap and (d) intermetallic compound.

composition the equilibrium phase or phases at a given temperature are those which give the minimum overall free energy at that temperature.† The difference between the free energies of the component phases before and after transformation is the driving force.

In many cases, the final equilibrium state is not reached and the system rests in a state of metastable equilibrium.

2.2. Nucleation

The classical theory of nucleation was developed by Volmer and Weber,[3] and Becker and Döring[4] for the condensation of a pure vapour to form a liquid. Subsequent theories for other transformations are based upon this earlier work. The theory considered homogeneous nucleation, that is the formation of one phase by the aggregation of components of another phase without change of composition and without being influenced by impurities or external surfaces. Impurity particles and external surfaces are taken into account in heterogeneous nucleation theory. The modifications to the classical theory which are necessary when examining vapour–solid, liquid–solid and solid–solid transformations must then be considered, as must the effects of compositional changes. Finally, we will study heterogeneous nucleation.

2.2.1. The Volmer–Weber–Becker–Döring theory (V–W–B–D)

Consider the free energy changes which occur if a spherical embryo of liquid is formed within a uniform vapour. First, there will be a change in free energy associated with the difference in volume free energy of the atoms in the liquid and the vapour.‡ Secondly, there will be a term introduced because a number of the atoms occur in the transition region between vapour and liquid. These atoms will be in a high energy state and are the origin of the surface free energy of the embryo.

For a spherical embryo of radius, r, the overall change in free energy, ΔG, is given by

$$\Delta G = 4\pi r^2 \gamma_{LV} + (4/3)\,\pi r^3 \cdot \Delta G_v \qquad (4)$$

where γ_{LV} is the surface free energy (ergs/cm^2) and

$$\Delta G_v\,(= G_{\text{liquid}} - G_{\text{vapour}})$$

† A full description of the free energy-composition diagram and its relation to the equilibrium phase diagram can be found in Cottrell[1] and Darken and Gurry[2].

‡ One of the inherent difficulties of nucleation theory is associated with the use of macroscopic thermodynamic properties, *e.g.* volume free energy and surface energy, in microscopic situations. This is unavoidable but does not normally introduce serious numerical error.

is the volume free energy change (ergs/cm^3). For an unsaturated vapour ΔG_v is positive and for a saturated vapour $(T < T_b)$ ΔG_v is negative. The variation of free energy with radius is as shown in Fig. 3. Any embryos

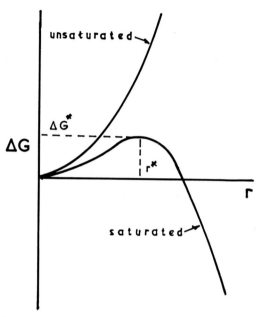

Fig. 3.   The change in free-energy due to the formation of a spherical embryo.

which form in an unsaturated vapour will thus rapidly disperse. On the other hand, for a saturated vapour provided the embryo reaches a critical size with radius r^*, at which $(\partial(\Delta G)/\partial r) = 0$, it is equally probable that it will disperse or that it will grow as a stable nucleus. To form this critical nucleus will require a random fluctuation producing a localised energy change, ΔG^*, given by

$$\Delta G^* = \frac{16\pi\gamma_{LV}^{\ 3}}{3(\Delta G_v)^2} \tag{5}$$

For the vapour–liquid transformation, assuming ideal gas behaviour, ΔG_v can be evaluated as

$$\Delta G_v = \frac{KT \ln i}{\Omega_L} \tag{6}$$

where K is Boltzmann's constant, T is the absolute temperature, Ω_L is the atomic volume in the liquid and i is a measure of the supersaturation namely

$$i = \frac{\text{saturated vapour pressure at temperature } T}{\text{vapour pressure in equilibrium with a flat}} \quad (7)$$
$$\text{liquid surface at temperature } T$$

Volmer and Weber's original theory[3] proposed a steady-state distribution of embryos which was the same as that obtained by assuming metastable equilibrium. The number of nuclei formed was determined by the rate at which an atom from the vapour could condense on embryos of a critical size. This led to a rate of nucleation, I, given by

$$I = Nq(4\pi r^*)^2 \exp\left(-\frac{\Delta G^*}{kT}\right) \quad (8)$$

where N is the total number of atoms, q is proportional to $(\rho/T)^{\frac{1}{2}}$ (where $\rho = $ vapour pressure) and ΔG^* is proportional to $1/(\ln i)^2$. This latter dependence means that the rate of nucleation will be extremely sensitive to the supersaturation. This treatment implies that no embryos exist with $r > r^*$.

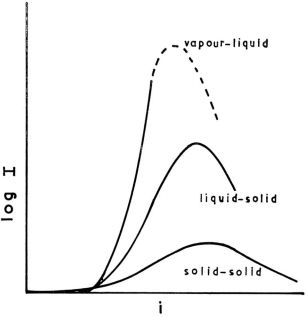

Fig. 4.   The dependence of the rate of nucleation on the degree of supersaturation (schematic).

This theory was modified by Becker and Döring[4] to allow for the dissolution of embryos of greater than the critical size. Their modification resulted in a change in the pre-exponential factor and a decrease in the nucleation rate by a factor of about 100.

Subsequently other efforts have been made to treat the problem more rigorously (*see* Hirth and Pound[5] and Christian[6]). In general these led to modified forms of the pre-exponential factor but produced no significant change in the nucleation rate.

Thus in all cases we can consider that there exists a temperature at which the rate of nucleation shows a sudden marked increase. This is as shown in Fig. 4 and as described earlier it is critically dependent on the supersaturation.

It also follows that the rate of nucleation will be dependent on the size of the system under examination. This result is common to all theories of homogeneous nucleation.

2.3.   The vapour–solid transformation

The V–W–B–D theory assumed the development of spherical embryos with isotropic surface energy. When we consider the nucleation of solids from the vapour phase we would anticipate the necessity of modifying the theory to take into account crystallographic constraints. Accepting that in general the surface free energy of a crystalline solid is markedly orientation dependent we would expect our embryos and nuclei to be of that form which minimised the overall surface energy per unit volume. This will be the shape determined by the application of Wulff's theorem.[7] Further modification will be necessary to allow for transient changes in shape and energy as we add atoms. Both contingencies have been allowed for in the several theories for this transformation.[5,6] The results are qualitatively in agreement with the V–W–B–D theory.

2.4.   The liquid–solid transformation

For this transformation we would expect the controlling equations to be similar in form to those for transformations from the vapour phase. However, it will be necessary to incorporate a term which is related to the mobility in the liquid state since this determines the availability of atoms. The treatment of Turnbull and Fisher[8] which followed that for vapour–liquid nucleation gave a rate of nucleation, I,

$$I = \frac{NkT}{h} \exp\left(-\frac{Q}{kT}\right) \exp\left[-\frac{16\pi\gamma_{LC}{}^3 T_m{}^2}{3L^2(\Delta T)^2 kT}\right] \qquad (9)$$

where h is Planck's constant, Q is the activation energy for transfer of an atom from liquid to solid, γ_{LC} is the solid–liquid surface energy, T_m is the equilibrium melting temperature, L is the latent heat and ΔT is the undercooling. This is analogous to the original V–W–B–D equations with ΔG_v evaluated as

$$\Delta G_v = \frac{L \cdot \Delta T}{T_m} \tag{10}$$

and including a pre-exponential term to compensate for the reduced mobility.

Substitution of acceptable values for the terms in the equation for nucleation rate indicates that undercoolings of $\sim 0.2T_m$ are to be expected for the homogeneous nucleation of solids from liquids. This has been verified experimentally for many metals.[9] At this undercooling the critical radius is $\sim 10^{-7}$ cm and the nucleus would contain approximately 200 atoms.

Homogeneous nucleation in alloys is a more complex process since the phase diagram requires equilibrium between a solid nucleus and liquid of different compositions. In this situation diffusional processes in the liquid will play an important part. There is at present no theoretical treatment for this problem. However, the experimental observations by Walker[10] of copper–nickel alloys showed that the undercooling required for homogeneous nucleation was ~ 0.2 of the liquidus temperature.

The form of the nucleation rate equation is such that at very large undercoolings there should be a decrease in the rate of nucleation resulting from a decreased mobility of atoms. We might thus expect the curve of Fig. 4 to show a maximum as indicated. For most metallic systems, however, it is not possible to cool through the range of temperature at which copious nucleation occurs at a rate fast enough to suppress nucleation. Quench rates of the order of 10^6 °C sec^{-1} have been used unsuccessfully with melts of pure liquid metals. In binary alloy systems where redistribution of solute must occur as part of the nucleation process some amorphous solids have been produced using rapid quenching techniques.[11,12,13]

The occurrence of amorphous phases results from the combined influence of two exponential factors, one related to the transfer of atoms from the liquid to the solid and the other associated with bulk diffusion in the liquid. On reheating, transformation to the crystalline state occurs quite rapidly, as shown in Fig. 5.

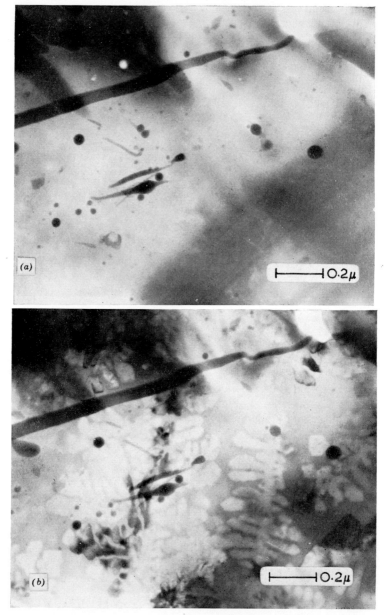

Fig. 5. The electron microstructure of splat-cooled tellurium −15 at. % germanium in (*a*) the amorphous state, and (*b*) after heating. The growth of dendrites can be seen quite clearly in this latter micrograph (Willens).[13]

2.4.1. *Non-metallic materials*

In non-metallic systems, *e.g.* ceramics and polymers, the suppression of nucleation and the production of amorphous structures can be achieved with comparative ease. The nucleation phenomenon in these systems is governed by the equations cited previously, however, because of the larger structural units which must be successively added to a crystal embryo before it can attain a critical size, there is much-reduced nucleation rate. The network structures occurring in silicates and other glass-forming oxides and in polymers, require large numbers of atoms to be located at the same time. This is difficult and so the term $\exp(-Q/kT)$ has a very dominant effect because of the increased activation energy required to

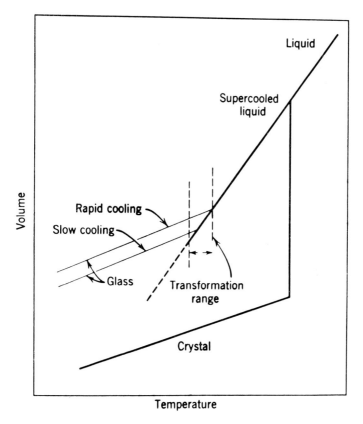

Fig. 6. Schematic representation of the origin of the glassy, or amorphous state.

orient the added unit. This can effectively suppress the region of rapid nucleation and allow the liquid structure to supercool to low temperatures. If nucleation occurs on cooling the liquid melt a discontinuous change in volume is observed. Even in the absence of nucleation a discontinuity in the expansion coefficient is normally observed and this is associated with the formation of a glass (Fig. 6). The temperature at which this transition takes place (the *fictive* temperature) is dependent upon the cooling rate and below this temperature the structure is amorphous and stable. Heating can produce crystallisation but any transformation takes place sluggishly. Controlled nucleation in glasses is used in ceramic technology to produce combinations of properties which are not feasible by other methods.[14]

2.4.2. *Heterogeneous nucleation*

In practice the majority of nucleation phenomena take place at undercoolings significantly less than those predicted by theories of homogeneous nucleation. For example, whereas undercoolings of the order of $0.2T_m$ ($\sim 200°C$ for most metals) would be expected in solidifying pure metals it is found experimentally that most metal melts nucleate at temperatures, only tens of degrees below the melting point. This discrepancy is attributed to the presence of a suitable surface in contact with the liquid. The nucleation is considered to be heterogeneous and to take place on the surface of the container or on particles present in the system. Heterogeneous nucleation can occur with essentially all transformations provided that some preferential sites exist.

The theory of heterogeneous nucleation has been developed for the vapour–liquid, vapour–solid, liquid–solid and solid–solid transformations. The theories for the first two of these transformations have been dealt with very fully by Hirth and Pound[5] and for the present purpose it would seem most useful to give detailed consideration to the theory for the latter two transformations only since these are most frequently involved in the development of microstructures in metallic and non-metallic materials.

For the simplest case of heterogeneous nucleation[15] in which a spherical cap of crystal forms on a planar substrate (Fig. 7) it can be shown, following a procedure similar to that of the theory of homogeneous nucleation, that

$$\Delta G^* = \frac{4\pi\gamma_{LC}{}^3(2 - 3\cos\theta + \cos^3\theta)}{3\Delta G_v{}^2} \tag{11}$$

where θ is the contact angle and the other symbols have their previous meanings. This differs from the value obtained in considering homogeneous nucleation by the factor

$$\tfrac{1}{4}(2 - 3\cos\theta + \cos^3\theta)$$

For $\theta = 180°$ the energy fluctuation required is the same as for homogeneous nucleation but for all other cases in which $0 < \theta < 180°$, heterogeneous nucleation is a more energetically favourable process.

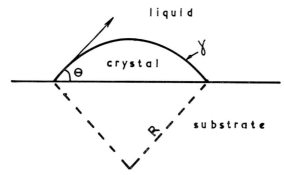

Fig. 7. Spherical cap of crystal formed on a planar substrate.

For systems in which the contact angle is small it is clear that the barrier to nucleation is also small and this can satisfactorily account for the low undercoolings observed in practice.

The theory can be modified to allow for different nucleus geometries, e.g. disc-shaped caps, without substantially affecting the overall result. If the substrate is allowed to have curvature or if the formation of nuclei in cavities on the surface is considered it is found that the barrier to nucleation is reduced even further.

The contact angle, θ, is a function of the surface energies of the liquid–crystal interface, γ_{LC}, the crystal substrate interface, γ_{CS}, and the liquid–substrate interface, γ_{LS}, namely

$$\cos\theta = \frac{\gamma_{LS} - \gamma_{CS}}{\gamma_{LC}} \tag{12}$$

For low contact angles it is desirable to have the crystal–substrate interface significantly less energetic than the liquid–substrate interface. A number of criteria have been postulated for this to be so. One of the most promising approaches was that of Turnbull and Vonnegut[16] which related the undercooling necessary for heterogeneous nucleation to the disregistry between the lattices of the crystal nucleus and the substrate at the interface. More recently Sundquist and Mondolfo[17,18] have questioned this approach and the evidence they put forward strongly suggests that the nucleation process is much more complex.

Nonetheless, it is well-established that the activation energy for heterogeneous nucleation, ΔG^*, is much less than that for homogeneous nucleation and proceeding as before it is possible to obtain an expression for the rate of nucleation that is similar in form to that for homogeneous nucleation. The smaller value of ΔG^* leads to higher rates of nucleation at smaller undercoolings and a much less sharp transition from low rates to high rates of nucleation. In addition since the heterogeneous nucleation process depends upon the presence of suitable sites, it is concluded that the rate of nucleation will pass through a maximum and show a cut off at higher undercoolings. This results from lateral spread of nuclei across the substrate and the consequent reduction in the surface area available for new nuclei. As with homogeneous nucleation it is not common for diffusion in the liquid to act as a limiting factor which reduces the nucleation rate at larger undercoolings. Figure 8 summarises schematically the relative difference between the two forms of nucleation.

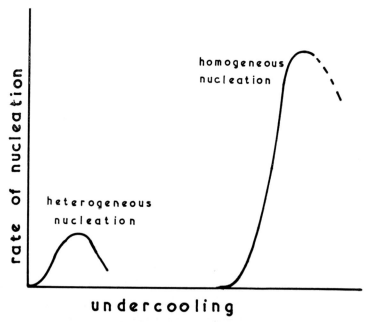

Fig. 8.   The relative rates of nucleation at decreasing temperatures for heterogeneous and homogeneous nucleation processes.

Recently attention has been paid to heterogeneous nucleation in alloy systems and in particular in eutectic alloys.[19] Although the general features of the nucleation process in this case are similar to those for pure metals one important difference concerns the strong dependence of the structure on the nucleation conditions.

In some cases, *e.g.* growth from dilute solutions, where the supply of atoms may exert a limiting effect it is common for surface imperfections such as emergent screw dislocations, grain boundaries or steps to act as preferred sites for heterogeneous nucleation. The subsequent growth structure usually reflects the influence of these modes of nucleation.

2.5. The solid–solid transformation

There are two major differences when the V–W–B–D theory is applied to solid–solid phase changes. Both of these have been alluded to in previous sections and one, the effect of a limitation in atom mobility has been shown to be important in the formation of glasses. However, the very much reduced diffusivity in the solid state (typically 10^{-10} cms^2 sec^{-1} near the melting point compared with 10^{-5} cm^2 sec^{-1} in the liquid) means that quenching past the maximum rate of nucleation in a solid state phase change (*see* Fig. 4) is relatively easy and requires quenching rates of only 10^{-1}–10^{-2} °C sec^{-1} compared with 10^6 °Csec^{-1} for a liquid–solid transformation. The other difference is quite specific to solid–solid transformations and concerns the magnitude of the PdV term in eqns. (3) and (4). While this is insignificantly small (and is therefore omitted) in liquid–solid transformations owing to the zero resistance to shear of a liquid, it is a very important term in a solid–solid transformation because any small change in volume resulting from the formation of the nucleus is resisted by the rigidity of the surrounding matrix. The energy change is, of course, positive and therefore tends to prevent the phase change occurring. The magnitude of the strain energy stored in the solid as a result of forming a nucleus of a second phase is not easy to calculate since it depends on the relative elastic moduli of the two phases, the shape of the nucleus and its orientation with respect to the matrix. A general model[20] leads to a term of the following form for a spherical nucleus:

$$\Delta G_{\text{strain}} = (4/3)\,\pi r^3 (\varepsilon_{11}{}^T)^2 E' \tag{13}$$

where r is the radius of the nucleus as before, $\varepsilon_{11}{}^T$ is the stress free linear transformation strain due to the formation of the nucleus and E' is a function of the elastic moduli of the two phases.† The two important

† E' approximates to $3E/2$ if both phases have the same Young's modulus E.

points about eqn. (13) are first that the mismatch $\varepsilon_{11}{}^T$ appears to the power of 2 which means that moderate misfits can have a large effect on the nucleation probability. Second, the energy varies as r^3 in the same way as the driving force for the phase change. The effect of strain energy is therefore in direct opposition to ΔG_v and the supersaturation required for a given rate of nucleation is necessarily increased. Thus eqn. (4) becomes

$$\Delta G = 4\pi r^2 \gamma_{CC} + (4/3)\,\pi r^3\,\Delta G_v + (4/3)\,\pi r^3 (\varepsilon_{11}{}^T)^2 E' \qquad (14)$$

and eqn. (5) becomes

$$\Delta G^* = \frac{16\pi\gamma_{CC}{}^3}{3(\Delta G_v + (\varepsilon_{11}{}^T)^2 E')^2} \qquad (15)$$

These modifications mean that the final equation for the nucleation rate of one solid within another solid (the analogue of eqn. (9) for the liquid–solid transformation) is:

$$I = I_o \exp\left(-\frac{Q}{kT}\right)\exp\left(-\frac{16\pi\gamma_{CC}{}^3}{3kT(\Delta G_v + (\varepsilon_{11}{}^T)^2 E')^2}\right) \qquad (16)$$

where I_o is the pre-exponential term which can be evaluated in the same way as described for the liquid–solid transformation.[6] As before Q is the activation energy for the transfer of an atom from the matrix to the nucleus. In most cases this is simply the activation energy for bulk solute diffusion, but occasionally a particular interface structure or composition means that the transfer of the atom across the interface is the rate-controlling step.

Equation (16) predicts a much lower value of I than eqn. (9) at all supersaturations (see Fig. 4), because of the larger value of Q in the solid state and the reduction of the driving force for the reaction by the inclusion of the strain energy term. In terms of undercooling, the second exponential only gives a sensible rate of nucleation at undercoolings of many hundreds of degrees for values of $\varepsilon_{11}{}^T \simeq 5\%$. But if the temperature is reduced to that extent, the first exponential has a small value owing to the lack of atom mobility and again nucleation is difficult. Hence it is rare to have *homogeneous* nucleation of the *equilibrium* phase in a solid–solid transformation.

2.5.1. Heterogeneous nucleation

As in the liquid–solid transformation, heterogeneous nucleation plays an important role in solid state reactions. Surfaces and impurity particles can again act as nucleation sites but because they represent such a small fraction of the total volume of the system, they are generally unimportant.

Instead, the major sites for heterogeneous nucleation are grain boundaries and dislocations.

Grain boundaries, which are essentially internal surfaces, can be treated in a very similar way to nucleation on surfaces in a liquid–solid transformation.[21] The angle θ (eqn. (12)) is then simply the contact angle for the precipitate nucleus at the grain boundary. In practice there are crystallographic effects arising from the fact that the nucleus tends to form with an orientation relationship to one of the two grains[22] and hence it makes boundaries of different types (*see* Section 4) and thus different contact angles with each grain. More complex nucleation models must then be used[23] but the main effect on eqn. (16) is an additional term involving θ which is similar to that shown for heterogeneous nucleation of the liquid–solid transformation in eqn. (11). Grain boundary nucleation is extremely important in solid–solid transformations because of the importance of the grain boundary in determining the properties of solids. Metallographically, however, it contributes a relatively insignificant number of nuclei towards the progress of the reaction since the number of grain boundary sites is such a small fraction of the total sites even for fine grained materials.[24]

Dislocations act as heterogeneous nucleation sites because of the interaction of the dislocation strain field and the strain field of the nucleus. The effect is not dissimilar to the interaction of solute atoms and dislocations[25] and leads to a binding energy between dislocations and the nucleus which acts to reduce the magnitude of ΔG_{strain} given in eqn. (13). There is a consequent reduction in ΔG and an increase in the nucleation rate. The first calculation of this interaction was made by Cahn[26] who assumed that the entire strain energy of the dislocation was removed from the system by the formation of a cylindrical nucleus parallel to the dislocation line. This model gives an indication of the potency of a dislocation line as a nucleation site in a solid–solid transformation, but takes insufficient account of the crystallographic relation between the nucleus and the matrix to give an accurate prediction of the results of nucleation on dislocations. Since the strain field of the nucleus is seldom spherically or even cylindrically symmetrical, it interacts with the dislocation strain field in a way which depends on the mutual orientations of the dislocation and the nucleus.[24,27] No treatment to take these details into account has yet been devised. Even a modest density of dislocations (*e.g.* 10^7 cms cms^{-3}) can have a large effect on the course of a reaction because of the large number of sites affected. Many systems transform entirely as a result of precipitate nucleation on dislocations.

2.5.2. Metastable phases

The difficulty in nucleating the equilibrium phase in a solid state reaction (except heterogeneously) leads to a phenomenon which is peculiar to solid–solid transformations: the formation of metastable or 'transition' phases. Since the value of the activation energy for nucleation (eqn. (15)) is a balance between surface and strain energies on the one hand and the driving force for the reaction on the other hand, it is clearly algebraicly possible to decrease the former and the latter in such a way as to cause an overall reduction in ΔG. Decreasing the driving force for the reaction is equivalent to the formation of a phase which is less stable than the equilibrium phase but which nevertheless leads to a reduction in the free energy of the system. If this phase possesses crystallographic character-istics which result in a closer fit with the matrix thus reducing the surface and strain energies, the algebraic reduction of ΔG mentioned above is clearly possible. This is the basis of the formation of metastable phases in solid–solid transformations: a phase which is not favoured energetically in an equilibrium state, may lead to the greatest reduction of free energy in finely divided form because of the large contribution of the surface free energy in this condition. In particular, in the very finely divided form typical of the formation of nuclei, it may lead to a reduction in the value of ΔG^* (eqn. (15)) and hence an increase in the nucleation rate. A typical metastable phase may have a value of ΔG_v of about half that of the equilibrium phase[26] but γ_{CC} may be reduced by an order of magnitude and $\varepsilon_{11}{}^T$ by a factor of 2–5.

There is one metastable phase which can be defined quite specifically. This is the coherent phase which approximately corresponds to the equilibrium phase in composition but retains the crystal structure of the matrix. Cahn[29] has shown that this phase is defined by a free energy function $\varphi(c)$ for a composition c given by

$$\varphi(c) = f'(c) + \frac{2\eta^2 E c^2}{(1-v)} \tag{17}$$

where $f'(c)$ is the Helmholtz free energy density of a homogeneous material of composition c, η is the variation in lattice parameter for unit com-position change and E and v are Young's modulus and Poisson's ratio respectively. The thermodynamic function $\varphi(c)$ can be handled in just the same way as more conventional quantities, *e.g.* the well-known common tangent rule can be applied to determine the limits of miscibility and from these a *coherent* phase diagram can be constructed, Fig. 9. As expected for a phase which is less stable than the equilibrium phase, the solubility in the matrix with respect to the coherent metastable phase is greater than with respect to the (normally non-coherent) equilibrium phase. The concept of the coherent phase diagram is of great importance in interpreting the microstructure of alloys and it is now accepted that the coherent

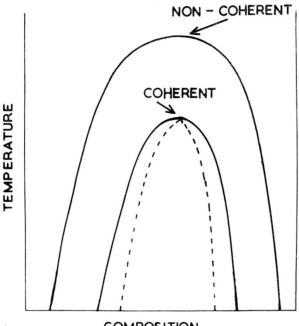

Fig. 9. A simple miscibility gap in a binary solid system. The non-coherent phase boundary refers to the equilibrium phase and the thermodynamic quantity f' (c), while the coherent phase refers to the metastable coherent phase and the thermodynamic quantity $\varphi(c)$. The dotted line is the coherent spinodal (Kelly and Nicholson).[27]

phase can be identified with the phenomenon of GP zones[30] which is discussed in Section 4.

The equilibrium phase and the metastable coherent phase do not represent the only possibilities in solid state reactions. Many systems show a series of transition phases of intermediate coherency which share the major property of the coherent phase that they result in a lowering of the value of ΔG^* (eqn. (15)) while being less stable than the equilibrium phase. Figure 10 shows an example of this for the Al–Cu system in which two further transition phases, θ'' and θ', appear between the coherent phase GP zones and the equilibrium phase θ (CuAl$_2$). This type of phase diagram is extremely useful but it must be borne in mind that the inclusion of non-equilibrium data on a phase (or equilibrium) diagram must be treated

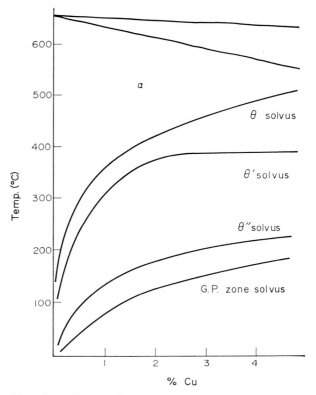

Fig. 10.   The phase diagram for the system aluminium-copper showing the limits of stability of the equilibrium phase, θ (CuAl$_2$), the metastable coherent phase (G.P. zones) and two metastable transition phases (θ' and θ''). (Drawn from Borelius *et al.*[32] and Beton and Rollason[28].)

with great caution. For example, the decomposition of an Al–Cu alloy within the θ'' phase boundary means that θ'' *may* form but it does not preclude the formation of the more stable phases θ' and θ.

2.5.3. *Spinodal decomposition*

As mentioned in Section 2.1, there is one class of transformation, spinodal decomposition, which can occur without a nucleation stage. The basis of this can be seen from Fig. 11 which shows a free energy/composition curve similar to Fig. 2(c). This diagram illustrates that a material of

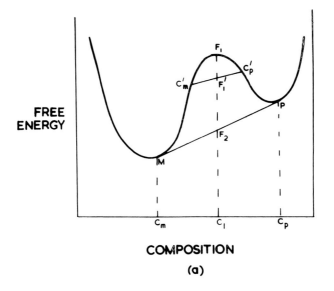

COMPOSITION

(a)

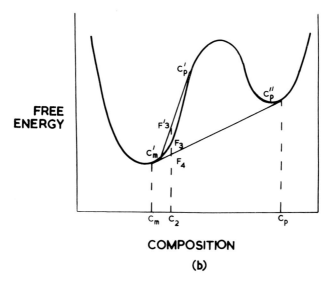

COMPOSITION

(b)

Fig. 11.   The variation of free energy with composition below the critical point of a system such as shown in Fig. 9, illustrating (a) spinodal decomposition, and (b) growth.

composition C_2 and free energy F_3 moves through a higher energy state F_3' at the start of decomposition and that a lower free energy, F_4, is only achieved when a substantial change in concentration occurs. This, of course, is a basic premise of nucleation theory: *i.e.* the nucleus has substantially the same composition as the final precipitate and is distinguished from the matrix by a sharp interface.

Consider now the more supersaturated alloys with compositions lying between the points of inflexion on the free energy/composition curve. An alloy of initial composition C_1 is able to decompose by a small amount so that the free energy decreases continuously from F_1 to F_1', and finally to F_2. Thus no nucleation barrier exists against the gradual perturbation of this supersaturated random solution towards complete phase separation. This process is known as spinodal decomposition.[30] Since the initial stages of this decomposition are small in degree, the fluctuations must be coherent with the matrix and hence it is the coherent phase diagram that we are concerned with.[29] For a fluctuation c' from a matrix of composition c:

$$\varphi(c, c') = f'(c) + \frac{2\eta^2 E(c - c')^2}{(1 - v)} \tag{18}$$

In making a free energy balance similar to eqns. (4) or (14), a term must be included which gives the extra energy associated with the diffuse interface between the fluctuation and the matrix. Cahn and Hilliard[31] proposed that:

$$F = \int_v [\varphi(c) + K(\nabla c)^2] \, \mathrm{d}V \tag{19}$$

where K is the gradient energy coefficient and V is the volume. Evaluation of eqn. (19) as a function of the supersaturation $\partial^2 \varphi / \partial c^2$ (or $\partial^2 f' / \partial c^2$ with respect to the equilibrium solvus) and the wavelength of the fluctuation λ (which is in effect the 'particle size') shows that the solution is unstable to fluctuations of wavelength greater than λ_c where

$$\lambda_c = \left(-\frac{K}{(\partial^2 \varphi / \partial c^2)} \right)^{\frac{1}{2}}$$

or

$$\lambda_c = \left(-\frac{K}{(\partial^2 f' / \partial c^2) + (2\eta^2 E/(1 - v))} \right)^{\frac{1}{2}} \tag{20}$$

This expression bears a marked resemblance to expressions for the critical radius r^* for a nucleation phenomenon derived from eqns. (4) or (14).†

† It should be noted that the relation between η and $\varepsilon_{11}{}^T$ is that $\eta(c - c') = \varepsilon_{11}{}^T$.

Equation (20) shows that spinodal decomposition can only occur when $\partial^2\varphi/\partial c^2$ is negative, *i.e.* within the points of inflexion on the free energy/ composition curve for the coherent phase—a result already deduced graphically from Fig. 11. It also shows that the critical wavelength is reduced for small gradient energies, large supersaturations and small misfits which is identical to the behaviour of r^*.

Most materials are elastically anisotropic so that $E/(1 - v)$ is a function of direction in a crystal. This means that λ_c is a function of direction and so is the rate of decomposition if an isotropic diffusivity is assumed. The consequences of this anisotropy will be considered in Section 4.

3. DEVELOPMENT OF MICROSTRUCTURE FROM THE LIQUID STATE

In the study of the development of microstructure from the liquid state logically we must first consider the growing nucleus. This is best done by examining the nature of the interface between the growing solid and the liquid. The structure and form of this interface influences both the microstructural morphology of the resultant solid and also the number and distribution of imperfections within the solid. It also has an effect on thermal and constitutional changes in the adjacent liquid and the interaction between these effects can in turn lead to growth modifications. The nature and extent of these modifications can be studied in the first instance by an examination of the behaviour of single phase materials during solidification. Subsequently, polyphase systems, *e.g.* eutectics, peritectics and monotectics must be examined.

It also follows that developments which can properly be called microstructural, for instance, control of grain structure, should also be considered since they are often a first means of influencing the transformation from liquid to solid, in a way that produces enhanced properties.

3.1. The structure of the interface

The interface can broadly be defined as the boundary between the liquid and the solid. It is normally described as smooth when the boundary is discrete and rough when the transition extends over a number of atomic layers. Jackson[33] following a suggestion of Burton, Cabrera and Frank[34] carried out an examination of the equilibrium structure of a solid interface in contact with a liquid. It was shown that starting with an atomically smooth surface and adding atoms at random the relative change in surface free energy, ΔFs, was given by

$$\frac{\Delta Fs}{NkT_E} = \alpha \cdot x(1 - x) + x \ln x + (1 - x) \ln (1 - x) \qquad (21)$$

where N is the number of possible sites on the interface, k is Boltzmann's constant, T_E is the equilibrium melting temperature, x is the fraction of sites occupied and

$$\alpha = \frac{L_o \xi}{kT_E} \qquad (22)$$

In eqn. (22) L_o is the latent heat of solidification and ξ is a crystallographic factor usually greater than 0·5. α depends on the material and the phase from which the crystal is growing. Figure 12 shows the expression of

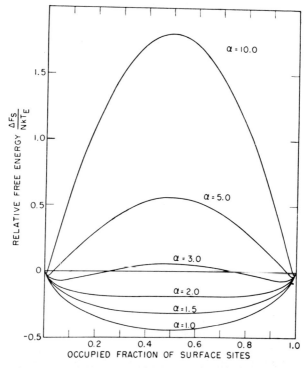

Fig. 12.   Relative change in surface free energy as a function of the fraction of surface sites which are occupied. α depends on the crystal face, the type of crystal and the phase from which the crystal is growing (Jackson).[33]

eqn. (21) plotted against x for various values of α. Two distinct types of interface can be deduced from this figure. First, for $\alpha \leq 2$ the interface has a minimum energy when approximately half the sites are occupied. On the other hand, for $\alpha \gtrsim 5$ the relative free energy is at a minimum when there are only a few occupied sites or a few unoccupied sites. The first type of interface was classified as 'rough' and the second type of interface as 'smooth' or 'facetted'. Further analysis showed that most metals had $\alpha < 2$, in which case they were expected to grow with a rough interface whose position was approximately determined by the isotherm fractionally below T_E. Inorganic and organic liquids normally have $\alpha > 5$ in which case the growing nucleus rapidly becomes bounded by crystallographic faces. A small group of materials exist, e.g. silicon, bismuth, which occupy the middle ground in that $\alpha = 2 - 5$ for different faces. In these materials the behaviour is more complex and often a mixed growth form results.[35,36] It must be emphasised that the analysis of Jackson considers the relative

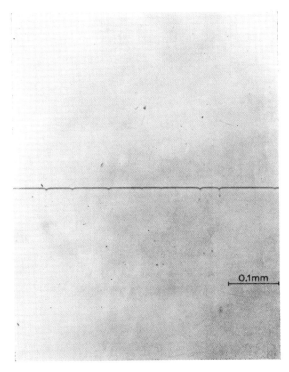

0.1mm

Fig. 13(a)

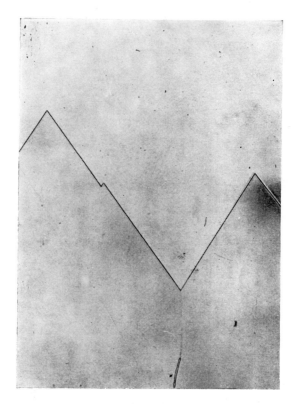

Fig. 13(*b*)

Fig. 13. (*a*) Planar 'rough' interface in carbon tetrabromide ($\alpha \sim 1$). The small depressions on the interface indicate the positions of grain boundaries. (*b*) Facetted interface in salol ($\alpha \sim 7$) (Jackson and Hunt).[38]

free energy change with the fraction of sites occupied on the surface and thus in essence it does not predict the true minimum energy form of a stationary equilibrated interface. This latter point has been examined by Miller and Chadwick.[37]

Nevertheless, the treatment has been of considerable value in increasing the understanding of the factors influencing the interface form during solidification. The studies of Jackson and Hunt[38] and Jackson, Uhlmann and Hunt[36] in which transparent organic crystals, chosen with a range of α factors and thus analogous to a range of materials both metallic and non-metallic, were observed during solidification, were largely in agreement with Jackson's original predictions.[33] Figure 13 shows examples of the different types of interface.

There are inadequacies in the theory of Jackson, particularly those associated with kinetic influences and related to details of the crystal structure, *e.g.* the anisotropy of growth.

One approach to the problem of the relation between the interface structure and the growth of the interface was that of Cahn[39] (*see also* Cahn, Hillig and Sears[40]) who considered in detail the 'diffuseness' of the interface, that is the number of atomic layers comprising the transition from solid to liquid. Cahn concluded that the degree of diffuseness was dependent both on the material and the driving force for transformation. The driving force is determined by the undercooling at the interface. As a result it was predicted that at low driving forces the interface would be discrete and propagation would take place by the transverse motion of interface steps, while at large driving forces the growth would be normal with the interface diffuse. This theory was examined in detail by Jackson, Uhlmann and Hunt[36] who compared the predictions with experimental data for a considerable range of materials. The examination showed that the evidence did not support the predictions.

It is thus apparent that further study is necessary to resolve these difficulties.

3.1.1. *Growth*

Notwithstanding the conflict over the nature of the interface, we must pay attention to the modes of crystal growth. A crystal grows when atoms (or molecules) are added to it. There are several different models which predict different dependences on the driving force. The normal approach is to consider the relationship between the mean rate of interface motion and the driving force.

In normal growth[41] all sites on the interface are considered to be equivalent and the interface advances by the continuous random addition of atoms. The growth rate is proportional to the undercooling ΔT. For normal growth the growth rates can be quite high and the requirement of site equivalence implies the need for a rough interface. Alternatively, if the interface advances using a step growth mechanism either involving a repeatable growth step, *e.g.* an emergent screw dislocation,[42] or by repeated surface nucleation,[43] the interface is facetted and the resultant growth rates are small. These two modes of growth have rate laws

$$R \propto (\Delta T)^2 \qquad (23)$$

for the growth on screw dislocations, and

$$R \propto \exp\left(-\frac{b}{\Delta T}\right) \qquad (24)$$

for repeated nucleation.

Apart from the rate control exerted by the method of growth, it is possible for other rate-controlling mechanisms to occur. These are principally (*i*) the rate of supply of atoms to the interface. This is normally only important in growth from the vapour or from dilute solutions although in the growth of alloys diffusional effects can be important, (*ii*) the rate of removal of latent heat. In contrast to (*i*) this is important for the solidification of metals and alloys but not for vapour growth or dilute solution growth.

3.2. Transformations in single phase materials

3.2.1. Pure materials

The growth of pure metals in a region of positive temperature gradient is controlled by the flow of heat away from the interface through the solid. The interface is normally rough and isothermal being at a temperature below the equilibrium temperature just necessary to provide sufficient kinetic driving force. This kinetic undercooling has been estimated as $\sim 0.01°K$. Non-metallic materials of high purity are expected to behave in a similar way within the constraints conferred upon them by the need to fulfil the interface requirements outlined in the previous section. The growing interface for both groups of materials should progress in a stable form. However, if a state of metastability or instability is created near the growing interface by the occurrence of an inverted gradient of free energy (this is normally achieved by inverting the temperature gradient) the growing interface will break down and grow dendritically.[44] In practice these conditions will only exist in bulk materials during the period between the initial nucleation and the attainment of the transformation temperature at which the dendrite arms thicken. Thus, if we consider a typical form of cooling curve it can be divided into sections as shown in Fig. 14. From a structural point of view the important parameters are those of grain size and shape. These are macrostructural variables and will be considered in Section 3.4. Of minor importance are the extent and nature of crystal defects, *e.g.* vacancies and dislocations. It is not expected that the concentrations of vacancies either singly or in clusters will be large under normal solidfication conditions,[45] nor that properties will be significantly affected. Dislocations, however, can have a significant influence. The grown-in dislocation density is determined by interactions during growth and by subsequent multiplication processes. With care, the overall densities can be kept low for metals[46] and reduced to zero for

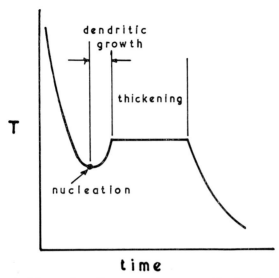

Fig. 14. Typical form of cooling curve for pure metals showing the regions in which the different growth phenomena occur.

non-metallic materials. Lineage structures, *i.e.* arrays of dislocation sub-boundaries,[47,48] also frequently occur but the available evidence indicates that they result from entrapped impurity particles.[49]

3.2.2. Alloys

The growth of impure metals is much more complex because of the redistribution of solute that occurs during solidification. This can produce changes in the growth morphology and lead to solute segregation on both a microscale and a macroscale. In the presence of a concentration of solute, C_o with distribution coefficient, k_o, it has been shown by Tiller *et al.*[50] that the equilibrium liquidus temperature ahead of an interface advancing at a rate, R, is given by

$$T_L = T_i + \frac{mC_o(1 - k_o)}{k_o}\left[1 - \exp\left(\frac{-Rx}{D}\right)\right] \qquad (25)$$

T_i is the interface temperature, m is the slope of the liquidus line for the specific system, D is the solute diffusivity and x is the distance ahead of the interface. The actual temperature is given by

$$T = T_i + Gx \qquad (26)$$

where G is the temperature gradient in the liquid ahead of the interface. As G varies a situation can arise where the liquid ahead of the interface is at a temperature below its equilibrium liquidus temperature, in which case a region of supercooling exists ahead of the interface. This is shown in Fig. 15. This phenomenon is known as constitutional supercooling. The

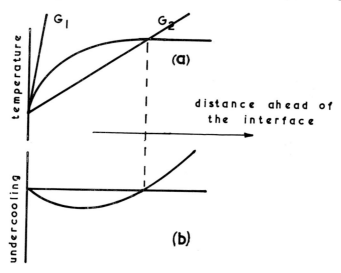

Fig. 15.   Constitutional supercooling ahead of an interface: (a) temperature distributions showing two applied temperature gradients, and (b) the magnitude of the supercooling.

limiting condition is where the temperature gradient is just tangential to the liquidus curve at the interface. Thus it can be shown that *no* constitutional supercooling will occur provided that,

$$\frac{G}{R} \geq \frac{mC_o}{D}\frac{(1 - k_o)}{k_o} \qquad (27)$$

In the absence of constitutional supercooling, the behaviour during growth is essentially the same as for pure materials with the exception that long-range segregational effects occur which are associated with initial and final transients in the solidification process (Fig. 16).

The above treatment assumes that solute mixing in the liquid is the result of diffusion only. It can be modified to allow for partial or complete mixing in the liquid (*see* Chalmers[51]). In these circumstances long-range segregation will still occur. This is also shown in Fig. 16.

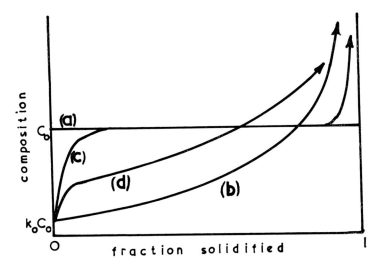

Fig. 16.   Solute distributions in a solid bar frozen from liquid of initial concentration, C_o, for (*a*) equilibrium freezing, (*b*) complete solute mixing in the liquid, (*c*) solute mixing in the liquid by diffusion only, and (*d*) partial solute mixing in the liquid.

The existence of a zone of constitutional supercooling, and thus a gradient of free energy, ahead of the interface will make an initially planar interface unstable to perturbations in shape. At low degrees of supercooling a cellular interface develops (Fig. 17). The breakdown from the planar-to-cellular form can be shown experimentally to be controlled by eqn. (27).[52,53] As the degree of supercooling increases the cells become more elongated and eventually cellular dendrites are formed (Fig. 18). There is no clear criterion for the interface transition from cells to cellular dendrites.[53] In the case of both cells and cellular dendrites, microsegregation occurs. The intercellular or interdendritic regions are rich in solute for $k_o < 1$ and depleted of solute for $k_o > 1$. Under these conditions long range segregation effects similar to those shown in Fig. 16 are also bound to occur.

On a macroscopic scale this means that a dendrite will show a variation

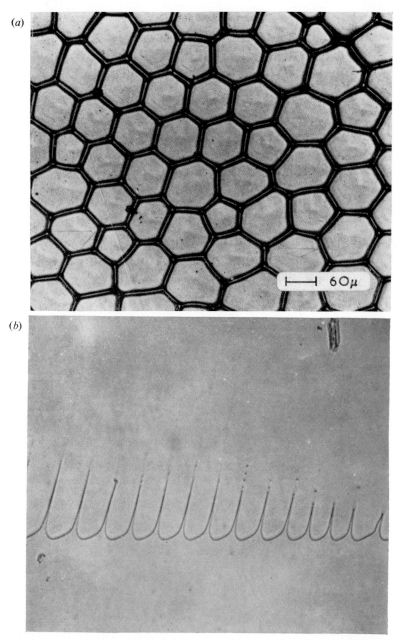

Fig. 17.   The cellular interface structure. (*a*) normal view of a decanted interface, and (*b*) view of growing cellular interface in impure carbon tetrabromide (Jackson and Hunt).[38]

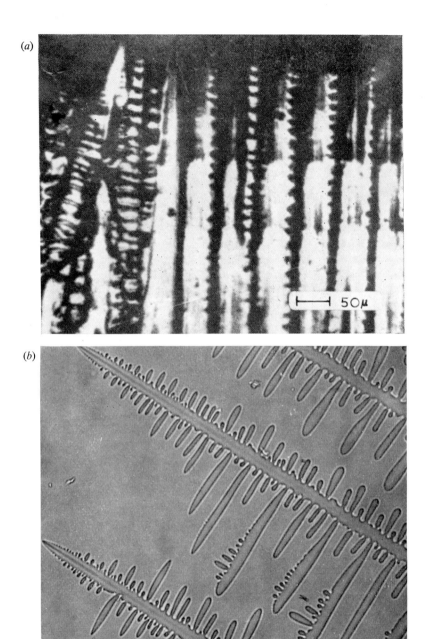

Fig. 18. The cellular-dendritic interface: (a) top surface of an impure lead alloy showing the development of branches on elongated cells (Tiller and Rutter),[52] and (b) dendritic growth in cyclohexanol (Jackson and Hunt).[38]

(a)

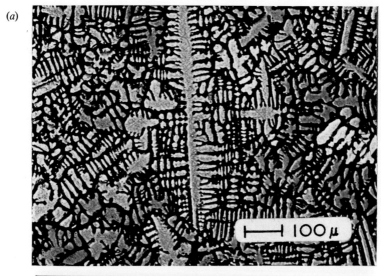

(b)

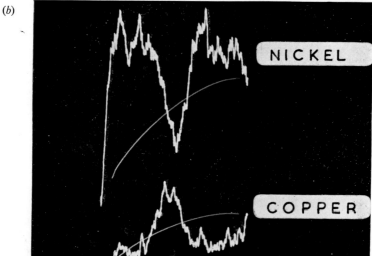

Fig. 19. Coring in chill-cast cupronickel ($k_o < 1$): (a) as-cast structure, and (b) electron-beam microanalyser trace across the boundary between two dendrite arms. The qualitative nature of the segregation is apparent as a maximum and a minimum in the copper and nickel, respectively.

in composition from inside to out; the phenomenon of *coring*. This is illustrated in Fig. 19. In this figure the dendritic structure is clearly evident. Although it cannot be resolved, the dendrite arms would undoubtedly show microscopic cellular segregation. Both forms of segregation can be eliminated by a homogenising heat treatment although, of course, macrosegregation effects involve long diffusion distances and thus long heat-treatment times. The effect of homogenisation is shown in Fig. 20.

A quite different approach to the problem of interface breakdown is that of Mullins and Sekerka,[54] who considered the stability of the interface in the presence of shape fluctuations. This treatment has some advantages, particularly since it predicts some of the parameters, *e.g.* cell size, which were not treated in the analysis of constitutional supercooling.

At this stage it would seem appropriate to point out that all the early studies of the interfaces of metals and alloys and the corresponding microstructural observations were handicapped by a basic experimental difficulty. This arose from the energetic constraints that ensured that a layer of liquid was retained when a solid–liquid interface was decanted. The recent direct observations of interface behaviour in organic analogues of metals[36,38] and in thin films solidifying in the electron microscope[55] have thus given valuable confirmation of the different theoretical mechanisms.

3.2.3. *Non-metallic materials*

In impure single-phase non-metallic materials the major differences in behaviour are the result of two factors:

(*i*) the general existence of facetted growth forms, and

(*ii*) the differences in atomic (or molecular) mobilities.

The former gives rise to cellular structures which are bounded by crystallographic faces in contrast to the well-rounded cells of metallic alloys. The latter leads to somewhat different segregational effects.

3.2.4. *Crystallographic effects*

Dendritic growth is strongly crystallographic. For example, it is found that dendrites of f.c.c. metals and alloys have their arms parallel to $\langle 001 \rangle$ directions.[44,56] When dendritic growth occurs continuously over a large volume of material the resultant structure is strongly anisotropic. This anisotropy can have quite deleterious effects on properties. For this reason steps are normally taken to promote profuse nucleation and ensure

(a)

(b)

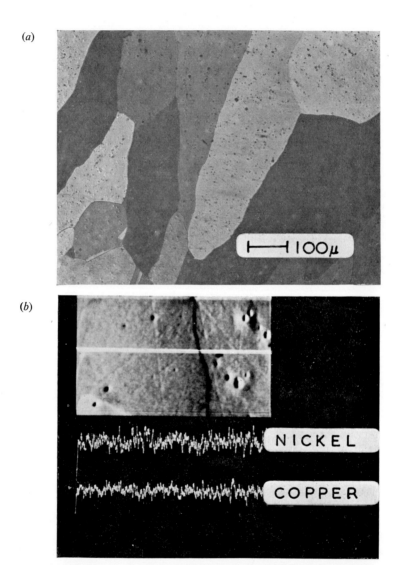

Fig. 20.   The structure of Fig. 19 after homogenising: (a) grain structure. The small dark particles are intermetallic impurities; (b) electron-beam microanalyser trace across a grain boundary. The boundary is the dark line in the electron image. The traces correspond to the white line in the electron image.

maximum isotropy. This important feature is considered further in Section 3.4.

3.3. Transformations in multiphase materials

3.3.1. *Eutectics*

In a pure binary eutectic mixture the two components solidify simultaneously to give a structure consisting of an intimate mixture of phases. Several different classifications for eutectic structures have been proposed[57,58,59,60] but there is not general agreement. The classification of Hunt and Jackson[60] has distinct advantages and is based upon the α-factor (eqn. (22)) characteristics of the component phases. Thus, the different structures are divided into three groups—(*i*) rough–rough, (*ii*) rough–facetted, and (*iii*) facetted–facetted—where the descriptions rough and facetted are as given in Section 3.1. Figure 21 shows examples of each of these groups for organic eutectics. Most metallic eutectics fall into group (*i*) and during growth there is coupling between the phases. The normal morphologies are lamellar or rod-like (Fig. 22). In the presence of impurities cellular growth occurs and a structure of eutectic colonies results[62] (Fig. 23). In lamellar eutectics the interlamellar spacing, λ, is related to the growth rate by

$$\lambda \propto \frac{1}{R^{\frac{1}{2}}} \qquad (28)$$

a relation predicted theoretically[63,64] and confirmed experimentally.[65,66] This provides one means of control of the eutectic structure.

For non-eutectic compositions the eutectic reaction is preceded by primary phase dendritic growth. The resulting structure consists of dendrites in a eutectic matrix (Fig. 24).

With group (*ii*) eutectics the growth process is still coupled with the facetted phase dominant. The resultant structures are less regular. As can be seen in Fig. 21(*b*) the facetted phase is surrounded by the other phase except at the tips of the growing spikes. The important metallurgical systems of aluminium–silicon and iron–graphite are of this type, as has been shown by Day and Hellawell.[67,68] Figure 25 shows the aluminium–silicon system. These two systems are both able to be structurally modified by the addition of small quantities ($\sim 0 \cdot 01$ per cent) of sodium and cerium or magnesium respectively. The recent work of Day[69] has shown that this modified structure is the result of a change in the mode of growth and not the result of a change in nucleation behaviour as was earlier proposed[70,71]; Fig. 26 shows examples of the modified structures

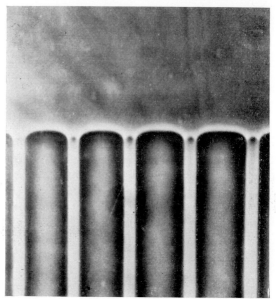

(*a*) A lamellar eutectic structure in the carbon tetrabromide-hexachlorethane system. Both phases normally have rough interfaces ($\alpha < 2$).

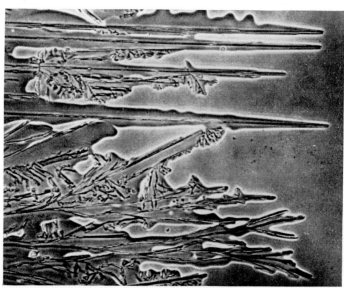

(*b*) An irregular eutetic structure in the succinonitrile (rough, $\alpha < 2$)—borneol (facetted $\alpha > 2$) system.

Fig. 21.

Fig. 21(c). Complex eutectic structure in the azobenzene-benzil system. Both phases normally have facetted interfaces. No coupled growth takes place (Hunt and Jackson).[60]

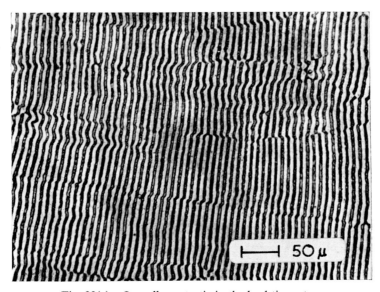

Fig. 22(a). Lamellar eutectic in the lead-tin system.

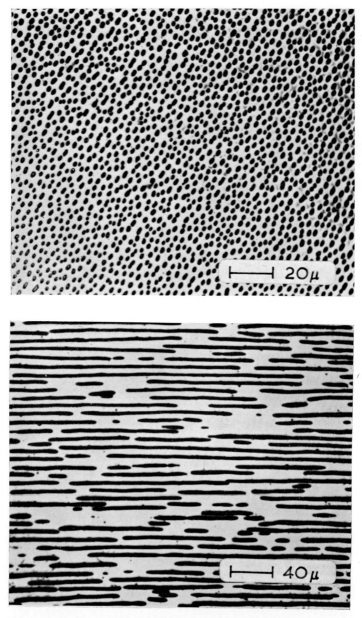

Fig. 22(b). Rod-like eutectic in the aluminium–Al$_3$Ni system: (i) transverse section. (ii) longitudinal section.

(a)

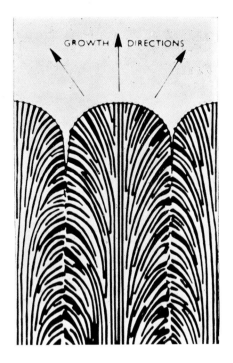

(b)

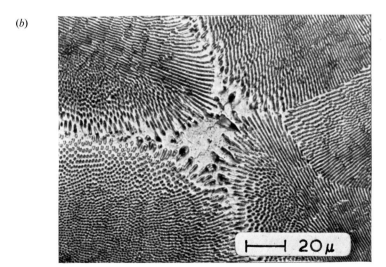

Fig. 23. (a) Schematic representation of the growth front of a lamellar eutectic growing with a curved cellular interface. Rods form at the cell boundaries (Hunt and Chilton).[61] (b) Cross-section through an impure lead–cadmium alloy showing the eutectic colony structure (Chadwick).[62]

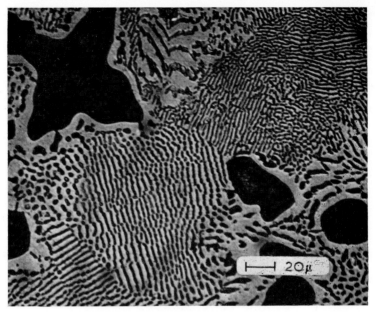

Fig. 24. Hypoeutectic alloy in the copper–silver system showing copper rich dendrites in a matrix of eutectic colonies.

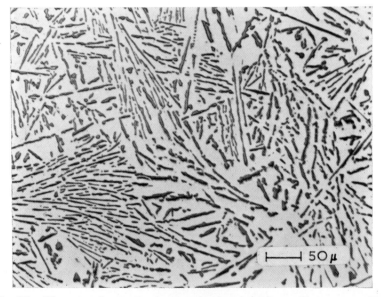

Fig. 25. The normal eutectic structure in the aluminium–silicon system (Day and Hellawell).[67]

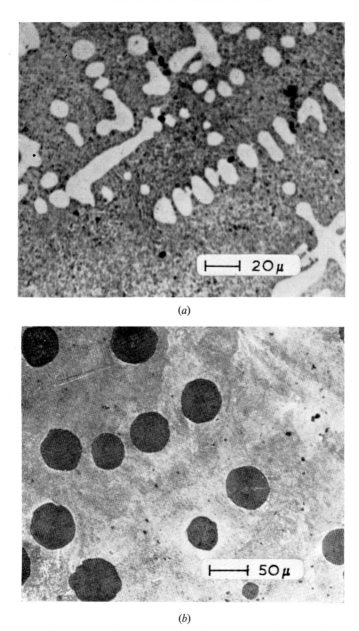

(a)

(b)

Fig. 26.   (a) Sodium-modified aluminium–silicon eutectic alloy, and (b) spheroidal
graphite cast iron.

of aluminium–silicon and iron–carbon (in the latter case the form of the graphite gives rise to the description 'spheroidal graphite iron'). Two points should be noted: (*i*) the modified structure has significantly improved mechanical properties, especially toughness, and (*ii*) the growth process is affected in such a way that primary dendritic growth occurs in what is nominally an alloy of eutectic composition.

These are both examples of important structural modifications in the development of microstructure.

Where both eutectic components are facetted the growth is no longer coupled and the resulting structure is a random mixture of the two phases (*see* Fig. 21(*c*)). This type of structure is uncommon in metallic systems but is expected for non-metallic systems where the α-factors of components should be large. This is not necessarily the case, however, since as shown by Hellawell and co-workers[72] inorganic eutectics, *e.g.* LiF–NaF, can grow with group (*i*) structures.

Ternary eutectic structures have not attracted a great deal of attention although some data are available.[73,74] Microstructures have been reported which have complex lamellar or combined lamellar-rod-like morphologies. Examples are shown in Fig. 27.

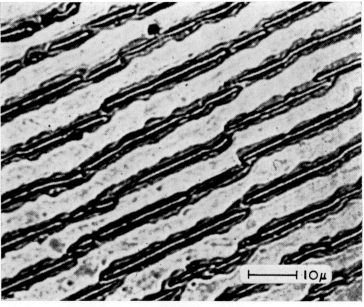

Fig. 27(*a*)

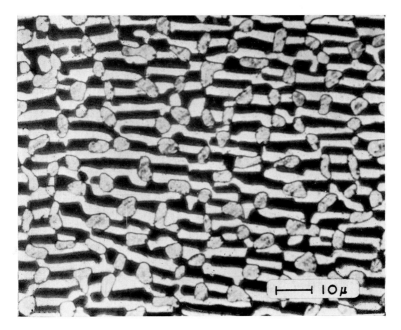

Fig. 27(*b*)

Fig. 27. (*a*) A triple lamellar ternary eutectic in the system cadmium (α) — tin (β) — lead (γ). The lamellar arrangement is $\alpha\gamma\beta\gamma\alpha\gamma$ (Kerr, Plumtree and Wine-gard).[73] (*b*) The complex ternary eutectic in the aluminium–copper–magnesium system. The lamellar phases are aluminium and $CuAl_2$ and the fibrous phase is Mg-rich (Cooksey and Hellawell).[74]

Structurally altered eutectic–dendritic structures have been grown by Mollard and Flemings.[75] In these alloys it was found possible to suppress the growth of primary dendrites over quite a wide range of compositions by control of the solidification conditions. Other eutectic systems of interest in the development of microstructure are the divorced eutectics[76] and the pseudo-binary eutectics[77] which occur in multicomponent systems. These are both very specialised topics.

Before considering other multiphase reactions we should note that eutectic systems have been examined closely by a number of workers with the specific intention of carefully controlling the development of the eutectic microstructure to give a planned anisotropy. The intention has been to produce fibrous composites by growth from the melt (*see* Hellawell[78] and Davies[79]).

3.3.2. *Peritectics*

Figure 28 gives the phase diagram for the almost ideal peritectic system, silver–platinum. As discussed by Uhlmann and Chadwick,[80] in the first detailed study of peritectic reactions, for virtually all compositions except those near the limits of the system, the microstructure after casting will consist of cored dendrites of one phase surrounded by the second phase. Thus with reference to Fig. 28, we would expect this structure for compo-

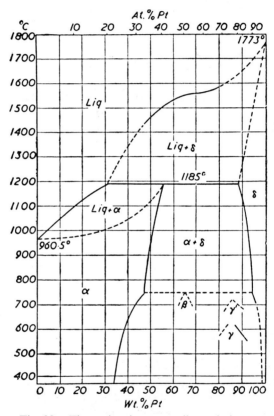

Fig. 28.   The peritectic system silver–platinum.

sitions from 30 to 90 wt per cent platinum. Furthermore, the peritectic reaction should seldom proceed to completion. When the primary phase has cooled to the peritectic temperature (1185°C in Fig. 28) it should react with the liquid to form the second component phase. In practice it becomes quickly encapsulated by the second phase and the reaction is stifled by the need for diffusion through this solid layer. This behaviour has been confirmed by Sartell and Mack.[81]

In those cases where successive peritectic reactions occur, e.g. tin-rich copper–tin alloys (see Fig. 29) considerable deviations from equilibrium

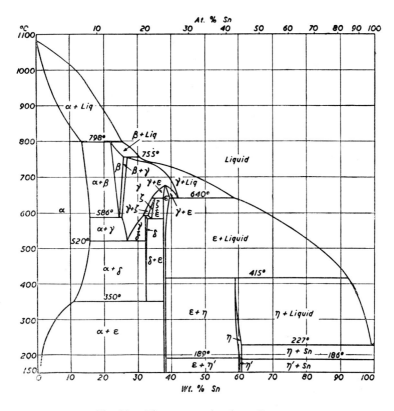

Fig. 29. The copper–tin phase diagram.

result. For instance if an alloy of nominal composition 80 wt per cent tin was solidified under equilibrium conditions, the resulting structure would consist of approximately equal proportions of the primary η–Cu_6Sn_5 phase and eutectic. When cast normally, however, it consists of the ε—Cu_3Sn phase enclosed in η–Cu_6Sn_5 in a matrix of eutectic. This is shown in Fig. 30.

On the whole, data for metallic systems are limited and there are essentially no data available for non-metallic systems.

Fig. 30.   As-cast copper—80 wt per cent tin. The intermetallic phases are $\varepsilon - Cu_3Sn$ (dark) in $\eta - Cu_6Sn_5$ (white) in a tin-rich eutectic matrix (black).

3.3.3. *Monotectics*

Monotectic reactions in which a liquid phase decomposes to form a solid and another liquid phase (Fig. 31) have also received little attention. The experimental work of Delves[82] and the theoretical treatment of Chadwick[83] showed that either regular rod-like structures or macroscopic phase separation should occur depending upon the various interphase surface energies. Subsequently Livingston and Cline[84] working with copper–lead alloys, established that the growth conditions also had a marked influence on the nature of the transformation structure. Their results are summarised in Fig. 32.

It would be valuable if these studies could be extended to other suitable systems.

3.4.   The control of the macrostructure during solidification

The control of grain structure is one of the most useful techniques available to the metallurgist or materials scientist who is concerned with the development of microstructure with the specific intention of improving properties. Many materials properties, especially mechanical properties, depend on grain shape and on grain size.

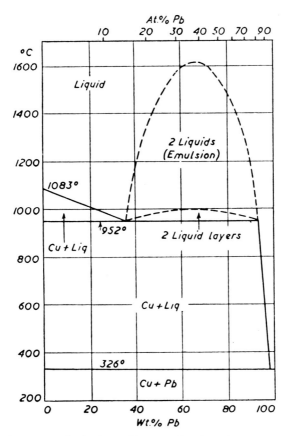

Fig. 31. The copper–lead phase diagram showing the monotectic reaction
$L_1 \rightarrow \alpha + L_2$.

(a)

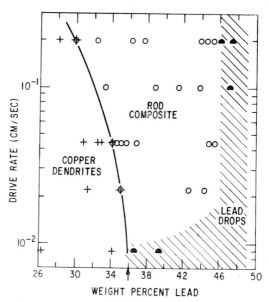

(b)

Fig. 32. (a) Composition-velocity plot showing regions over which different microstructure types were observed. (b) Transverse section of transition regions showing composite and dendritic structures copper-light phase: lead-dark phase. The characteristic alignment of the dendritic structure resulting from the crystallography of dendritic growth is clear (Livingston and Cline).[84]

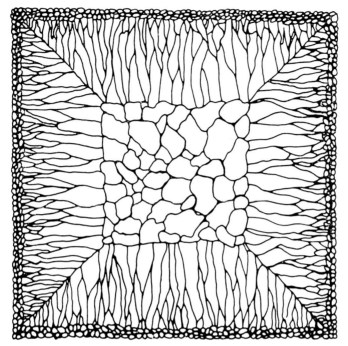

Fig. 33. Transverse section of an as-cast structure showing the chill zone, columnar zone and equiaxed zone.

In most as-cast polycrystalline solids three distinct zones with different grain structures can be identified (Fig. 33) namely:

(*i*) The chill zone—a boundary layer of small equiaxed crystals with random orientations.

(*ii*) The columnar zone—a band of elongated crystals which are aligned parallel to the directions of heat flow. These crystals show a strong preferred crystallographic orientation and are the result of directional dendritic growth.

(*iii*) The equiaxed zone—a central region of uniform crystals. The properties of this central region are comparatively isotropic provided the grain size is small.

The relative proportions of the different zones can be controlled by altering the casting variables, *e.g.* the alloy composition, the pouring temperature, the rate of cooling, etc. (*see*, for instance, Chalmers[85]).

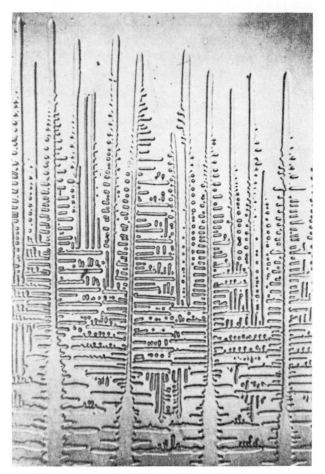

Fig. 34. Dendrite structure in carbon tetrabromide with salol added, after a growth rate fluctuation. The detachment by melting of dendrite arms can be clearly seen (Jackson *et al.*).[87]

The grain boundaries form by the impingement of the growing grains and the effects of the boundaries derive both from crystallographic sources and particularly because of the accumulation of soluble and insoluble impurities that can occur.

If the object of the control of grain structure is to obtain isotropy then a fine-grained equiaxed structure is required. This is brought about by encouraging those conditions which lead to the breakdown of columnar growth and the formation of the equiaxed zone. Understanding of the

Fig. 35(a)

Fig. 35(b)

Fig. 35(c)

Fig. 35. Successive stages in the solidification of an NH_4Cl–water ingot saturated at 50°C, poured at 75°C. (*a*) 1 minute after pouring. (*b*) 2 minutes after pouring. (*c*) 2½ minutes after pouring (Jackson *et al.*).[87]

factors that are responsible for the formation of the equiaxed zone has significantly increased in recent years. Chalmers[86] proposed a theory in which the nuclei that became the equiaxed zone were formed during the initial chill and swept to the centre where they eventually impeded columnar growth. Jackson and co-workers[87] on the other hand, argued that the equiaxed zone developed as a result of crystal multiplication brought about by the melting off of arms of the columnar dendrites, as shown in Fig. 34. This melting was produced by variations in growth rate associated with convective stirring. Certainly it is known that strong convection occurs in solidifying ingots and that this has a grain refining action.[88,89] However, other forms of crystal multiplication can occur because of mechanical effects. Certainly, the grain structure can be very much refined by the use of electromagnetically augmented convection,[90] dynamic nucleation by vibration,[91] or with ultrasonic pulses[92] (*see also* Chalmers[93]), or by vigorous mechanical stirring.[94] It is most probable that in many systems a combination of mechanisms is operative depending upon the actual conditions existing during solidification. Figure 35 shows successive stages in the formation of the ingot zone structure in a metal analogue system. The refining action that can be achieved practically by crystal multiplication is illustrated in Fig. 36.

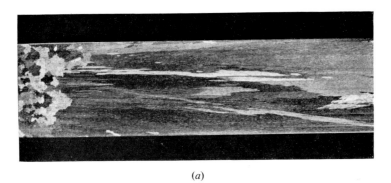

(a)

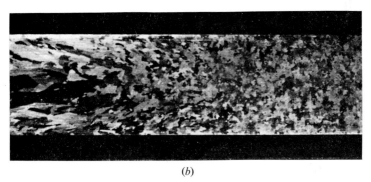

(b)

Fig. 36.   Macrostructures of aluminium–4 wt% copper solidified (a) without stirring, and (b) after stirring (Wojciechowski and Chalmers).[94]

It is also important not to overlook the refining action that can be brought about by the use of active heterogeneous nucleation.[95] In many instances this is a simple and effective means of gaining macrostructural control[96] (*see* Fig. 37).

More detailed microstructural control such as control of dendrite arm spacing, microsegregation, etc. is more difficult[89] but if the current rate of progress continues effective means of control should soon be available. As an example, Fig. 38, taken from the work of Flemings[89] shows a controlled transformation in magnesium–zinc alloys from coarse dendritic growth to spherical 'non-dendritic' growth. The transformation was brought about by the addition of zirconium.

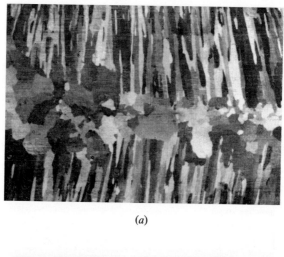

(a)

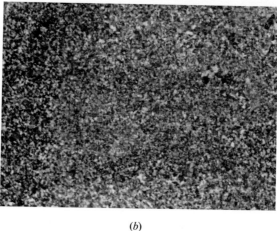

(b)

Fig. 37. (a) Normal coarse-grained cast structure in 18% chromium–12% nickel steel: (b) same steel as (a) refined by inoculation (Hall and Jackson).[96]

In conclusion, it should be noted that if the ingot structure is going to be subjected to extensive working in the solid state then the grain structure will be dictated by solid-state changes. Nevertheless, careful control in the initial stage can have beneficial results, particularly in so far as the attainment of homogeneity is concerned.

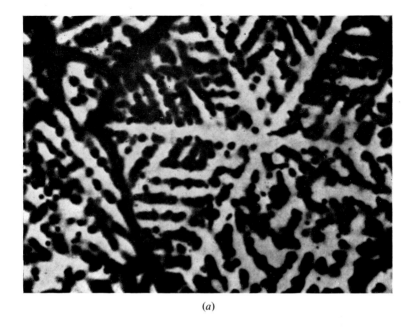

(a)

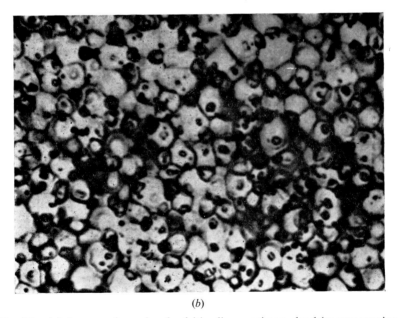

(b)

Fig. 38.　(a) A magnesium–zinc dendritic alloy specimen, dendrite arm spacing 40μ; (b) magnesium–zinc with zirconium grain refinement. Non-dendritic alloy specimen, grain size 70μ (Flemings).[8][9]

4. DEVELOPMENT OF MICROSTRUCTURE FROM THE SOLID STATE

4.1. Introduction

In this section we shall deal with microstructures arising from phase transformations occurring entirely in the solid state. One important category has already been considered in Chapter 4: the martensite transformation in which the atoms move from the parent to the product phases in a single co-ordinated movement at the interphase interface. Naturally this requires that the chemical composition of the two phases shall be the same. There remains the much larger category where the compositions of the two phases are, in general, different and where the atoms move from one phase to another in a variety of unco-ordinated movements so that the transformation is atomically irreversible. These transformations have been variously called 'nucleation and growth' which is a poor name since some of them do not involve a nucleation step, 'civilian' where the disorderly movements of the atoms are contrasted with the 'military' precision and co-ordination of atoms in a martensitic transformation,[6,97] and 'diffusion-controlled' which is probably the best name provided it is realised that one of many types of diffusion processes may be rate controlling.

4.1.1. *Interphase interfaces*

It will be clear from this introduction that the nature of the interphase interface is as important in the case of diffusion-controlled transformations as it is for the martensitic transformations considered in Chapter 4. A phenomenological description[27] of the possible types of interface will suffice although more rigorous crystallographic definitions are possible.[6] A coherent interface is one in which the plane of atoms constituting the interface is common to the crystal structures of both phases if chemical species is disregarded. This normally means that prominent planes of atoms run continuously from one phase to another. Such an interface can be of infinite length if the two crystal structures have identical lattice parameters; more usually there is a difference between the two lattice parameters and the interface region contains elastic strains which increase as the length of the interface increases until continuity or coherency finally breaks down and a semi-coherent interface is formed. A semi-coherent interface is one in which the crystal structures of the two phases are so similar that large parts of the interface are effectively coherent and occasional interface dislocations accommodate the misfit between the two lattices (Fig. 39). A non-coherent interface occurs between two phases when the crystal structures are so different that no simple fit between the two lattices is possible. In terms of interfaces, a coherent interface is equivalent to a coherent twin interface while semi-coherent and non-coherent interfaces are analogous to low angle and high angle grain boundaries respectively.

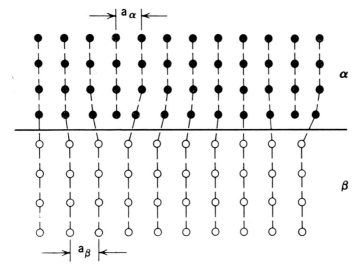

Fig. 39.   A semi-coherent interface between two phases α and β with slightly different lattice parameters (Turnbull).[98]

4.1.2. Energy of interfaces

The energy of an interphase interface is important in determining the microstructure resulting from a phase transformation. The energy can be expressed as the sum of two terms which are usually evaluated independently:

$$\gamma_{CC} = \gamma_{CC}^{\text{Geom.}} + \gamma_{CC}^{\text{Chem.}} \tag{29}$$

where $\gamma_{CC}^{\text{Geom.}}$ is the energy due to geometric effects at the interface, i.e. the strain energy, and $\gamma_{CC}^{\text{Chem.}}$ is the energy due to the specific arrangement (or order) of the atomic species at the interface. Expressions have been calculated for the enthalpic contributions to the two free energy terms[98–100] but no reliable method exists for calculating the entropic contribution. Typical values for $\gamma_{CC}^{\text{Chem.}}$ range from 10–250 ergs cm^{-2}, whereas $\gamma_{CC}^{\text{Geom.}}$ might be 50–200 ergs cm^{-2} for a semi-coherent interface and $\sim$350 ergs cm^{-2} for a non-coherent interface. The value of $\gamma_{CC}^{\text{Geom.}}$

is not readily defined for a coherent interface since the strain energy of an elastically strained particle is proportional to the volume of the particle rather than to its surface area.

It is impossible to generalise about the total energy of an interphase interface but it will be clear from the discussion above that the energies of different oriented interfaces between the same two phases can be widely different (*i.e.* the interface energy is anisotropic).

4.1.3. *Mobility of interfaces*

The mobility of an interface can be determined either by the rate of movement of atoms across the interface or by the rate of supply of the appropriate atom species to the interface. The former situation is typically found in allotropic phase changes in pure metals or in recrystallisation of single phase alloys. The velocity, G, of the interface then depends on the driving force for the reaction, ΔG_v, and the activation energy for diffusion across the interface Q_I:

$$G = K \frac{\Delta G_v}{T} \exp - \frac{Q_I}{kT} \qquad (30)$$

where K is a constant. This equation is valid for disordered or diffuse interfaces. If the interface is ordered and sharp, *e.g.* a coherent interface, the addition of an atom to cause a perturbation on an otherwise planar interface may require a substantial nucleation event. The problem is essentially a two-dimensional version of the nucleation situation detailed in Section 2. If growth is inhibited by a nucleation event, the normal growth of the interface will be inhibited and it can only migrate by the sideways movement of ledges on the interface. This situation is well known in crystal growth from the vapour.[6] It leads to very much smaller growth rates than predicted by eqn. (30) and also to facetted growth shapes bounded by well-ordered planes.

Whether or not growth takes place by normal movement of an interface depends on the driving force of the reaction and the diffuseness of the interface.[39] The situation seems fairly clear-cut for the vapour–solid case, controversial for liquid–solid changes and almost unexplored for the solid–solid situation.[101] This point is of considerable importance in discussing particle morphology.

If the new phase has a different composition from the matrix in which it is growing the supply of atoms of the right type by diffusion in the matrix may limit the rate of movement of the interface. The growth rate is then given by:

$$G = \frac{D}{c_\beta - c_\alpha} \frac{dc}{dx} \tag{31}$$

where D is the diffusivity, $c_\beta - c_\alpha$ the concentration difference between the precipitate and the matrix in equilibrium with each other and dc/dx is a measure of the concentration gradient in the matrix near the interface.

If the movement of an interface is determined by eqn. (31), the growth of a second phase is said to be 'diffusion-controlled'. If eqn. (30) or some other interface process is limiting, growth is said to be 'interface controlled'. Since so many solid state reactions concern a solute rich particle growing from a relatively dilute matrix and since the mobility of the solute atoms is often low, diffusion control is very common. In fact, there is a tendency for diffusion control to be assumed unless it is proved otherwise, but some recent experiments by Aaronson and co-workers[102,103] suggest that some form of interface control may be more frequent than is generally realised.

4.1.4. Morphology of the transformation product

The morphology of the transformation product of a solid state phase change is determined by one or other of two fundamental phenomena or occasionally by a mixture of the two. As in liquid–solid transformations discussed in Section 3, the morphology may be a true equilibrium one where the total free energy of the system is a minimum; alternatively, the morphology may be dictated by the kinetics of the situation with a non-equilibrium shape able to grow much faster than an equilibrium one for the reasons discussed above. Effectively these two extremes correspond to a maximum change and a maximum rate of change of free energy respectively. While the former should always be achieved in time, the rate of approach to equilibrium is often so slow that a kinetically determined morphology may appear to be an equilibrium one.

4.1.5. Continuous and discontinuous reactions

Diffusion controlled transformations may occur continuously or discontinuously and each process gives rises to particular microstructures. A continuous process is one which starts at a large number of centres within each grain of the parent material and proceeds at a rate which is approximately the same at any point in the material. Although the distribution of transformation product is inhomogeneous on a fine scale, it is homogeneous on a scale at which most physical properties are determined, *e.g.* the X-ray lattice parameter will show a steady change between the two extremes characteristic of the supersaturated and saturated matrix.

A discontinuous process is one where the reaction is wholly confined to one or more interfaces which sweep through the material. Thus, a partially transformed specimen is made up of some wholly transformed material and some material where the transformation has not started. Consequently, there will be no intermediate values of the X-ray lattice parameter as in a continuous reaction: the parameters of the initial and final solid solutions will be present throughout the transformation and only the intensities of the different reflections will change as the transformation progresses.

4.2. Precipitation from supersaturated solid solution

The rejection of a second phase from a solid solution which becomes supersaturated as the temperature is lowered is a frequent occurrence in the development of microstructure in the solid state. It gives rise to one of the oldest and best known of all microstructures, the Widmanstatten pattern on meteorites,[104] and is also responsible for the phenomenon of age-hardening which is presently the most widely used method of strengthening alloys.

With an equilibrium diagram like that shown in Fig. 10, the supersaturation of a solid solution will increase steadily as the temperature is lowered until a second phase is nucleated and starts to grow. In principle, equilibrium can be maintained as the temperature is reduced but, as discussed in Section 2, the nucleation frequency for a new phase is vanishingly small at small supersaturations and hence considerable undercooling is normally necessary before precipitation begins. Equilibrium is probably most closely approached in meteoritic structures where the cooling rate of $1°C$ per 10^6 years (effectively unattainable in terrestrial circumstances) allows phase separation to occur at very low supersaturations and correspondingly small nucleation frequencies. A typical Widmanstatten pattern is shown in Fig. 40(a) for the separation of the α phase from the γ phase in an Fe–Ni alloy. The coarse structure which is visible to the naked eye is a reflection of a nucleation density of the order of 1–10 particles per cc. Another relatively coarse structure is obtained from supersaturated Cu–Si alloys on cooling where plates of the κ phase are precipitated in the Cu rich matrix, Fig. 40(b).

Most Widmanstatten patterns are obtained by quenching and ageing so that the supersaturation at the start of ageing is high and the nucleation

Fig. 40(*a*)

Fig. 40(*b*)

262   STRUCTURAL CHARACTERISTICS OF MATERIALS

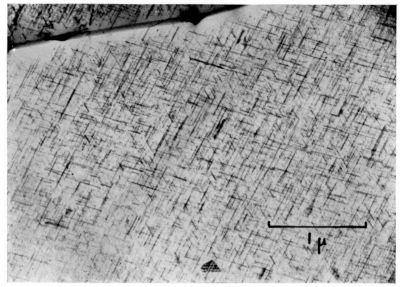

Fig. 40(c)

Fig. 40. Illustrating Widmanstatten patterns: (a) kamacite (α-iron) plates in taenite (γ-iron) in Laurens County meteorite (micrograph courtesy of F. Knowles); (b) κ plates (c.p.h. Cu_8Si) in α (f.c.c. Cu) in a Cu-Si alloy (Barrett and Massalski);[106] and (c) Cr_3P rods in γ-iron in a 18%Cr 10%Ni 0·3%P stainless steel (Rowcliffe).[107]

density is of the order of 10^{13}–10^{15} particles per cc. The microstructure is correspondingly finer than Fig. 40(a); an example of Cr_3P precipitates in an austenitic stainless steel is shown in Fig. 40(c). The formation of the plate shaped precipitates shown in Fig. 40 is always associated with the precipitation of a phase whose crystal structure is closely similar to that of the matrix over one crystal plane but not over the others, e.g. the formation of h.c.p. κ from f.c.c. α (Fig. 40(b)) where the basal plane of the κ structure and the {111} planes of the matrix are effectively identical and form a perfectly coherent interface. Thus four orientations of κ plates are formed with the plane of the plates parallel to each of the four {111} matrix planes. Amazingly there appears to be no general agreement as to whether the resulting morphology of Widmanstatten plates is due to kinetic or thermodynamic effects.[101] It can be argued that the difficulty of nucleating a new layer of atoms to thicken a Widmanstatten plate is due to the absence of heterogeneities on the smooth coherent or semi-coherent interface; hence the particle lengthens more rapidly than it thickens.[102,103] Alternatively,

as discussed earlier, the surface energy of the particle may be extremely
anisotropic so that the equilibrium shape is a plate or a needle rather than
a sphere as would be the case for a particle with an isotropic surface energy.
Currently the situation seems to be that the former explanation is accepted
for coarse Widmanstatten structures while the latter is generally assumed
to be true for fine structures[27] where the diffusion path along the interface
would appear to be too short for a non-equilibrium shape to be preserved
for long.

The Widmanstatten patterns shown in Fig. 40 are rareties in common
materials because they all concern the precipitation of the equilibrium
phase directly from the supersaturated solid solution. More often the
development of a microstructure to give useful properties means ageing
at a supersaturation where transition phases are formed, as described in
Section 2. Typically there is a sequence of precipitates formed as ageing
proceeds:

$$\alpha_{supersat} \rightarrow \alpha_{sat} + \text{G.P. Zones} \rightarrow \alpha_{sat} + \beta' \rightarrow \alpha_{sat} + \beta \qquad (32)$$

The composition of the saturated solid solution will, of course, decrease as
the precipitate sequence approaches the equilibrium phase, according

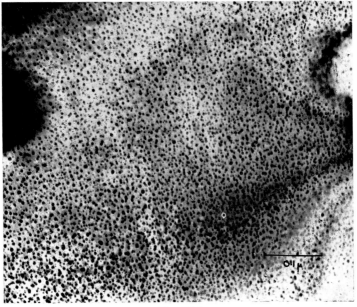

Fig. 41(a)

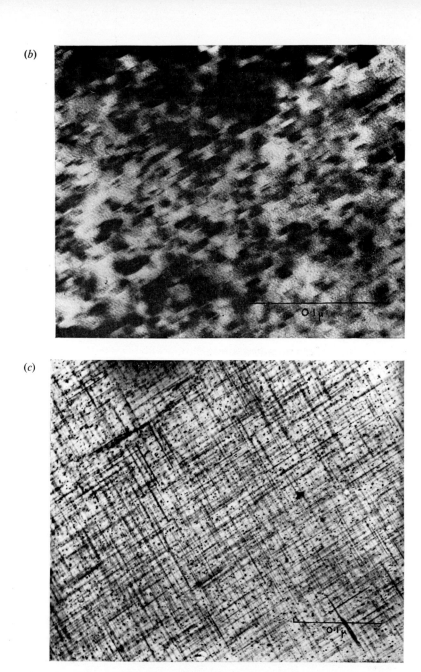

Fig. 41. Illustrating different shapes of G.P. zones in various Al alloys: (a) spherical zones in Al–Ag; (b) plate-shaped zones in Al–Cu; and (c) rod-shaped zones in Al–Cu–Mg (Weatherly).[117]

to the metastable solvus lines (*see*, for example, Fig. 10). G.P. or Guinier–Preston zones is a term which has traditionally been associated with the formation of small 'precipitates'.[108] The use of this term is open to serious objections[27] and, following Section 2 an acceptable definition of G.P. zones would be: the first formed transition precipitate providing this is completely coherent with the matrix. G.P. zones are therefore clusters of solute atoms having the same, or nearly the same, crystal structure as the matrix; they are distinguished from the heterophase fluctuations of nucleation theory (Section 2) by their permanence and specific composition.

G.P. zones may form as spheres, plates or needles depending primarily on whether there is any elastic misfit between the zone and the matrix and whether there is any order among the solute atoms. If neither of these effects is present the zones are spherical (Fig. 41(*a*)); if there is some order among the atoms, the zones may form a plate or needles[108] (*e.g.* Fig. 41(*c*)). Most commonly, however, G.P. zones take up a shape which minimises the elastic strain energy of the system. Although this depends only on the volume of the zone for elastically isotropic lattices, the energy can be reduced markedly by the formation of plates in elastically anisotropic lattices.† Since most cubic metals have a minimum value of the elastic constant along $\langle 100 \rangle$ directions, strained G.P. zones tend to form on $\{100\}$ planes in a variety of alloys, *e.g.* Fig. 41(*b*). The formation of G.P. zones appears to be an almost universal phenomenon in supersaturated alloys provided that the supersaturation is high enough so that the alloy is below the G.P. zone solvus temperature (*see* Fig. 10). G.P. zone formation is important in that some alloys achieve their maximum properties in this condition. In addition it has recently been recognised that the G.P. zone distribution can have a large effect on the distribution of the next transition phase by providing nucleating sites.[111–114]

The next stage in the ageing sequence is normally the formation of another transition phase which is partially coherent with the matrix,

---

† There is some confusion about this point in the literature. Nabarro[109] showed that a strained inclusion in an elastically *isotropic* lattice may take a plate or lath shaped form to minimise the strain energy. However, the elasticity theory used for this calculation assumes no mass transfer between the parent and product phases which is hardly likely to be the case for a diffusion controlled process with a vacancy mechanism operating. Therefore, although this result is widely quoted in the precipitation literature, it should really only apply to martensitic transformations where the product phase is indeed lath or plate shaped. In a second paper Nabarro[110] also showed that a coherent elastically strained precipitate would be plate shaped in an elastically *anisotropic* lattice and it is this result which is obviously relevant to precipitation phenomena since the model used is of more general application.

266   STRUCTURAL CHARACTERISTICS OF MATERIALS

i.e. one interface is coherent or semi-coherent while the other is non-coherent. This leads to considerable anisotropy of the surface energy and such phases typically have strongly geometric shapes: laths, discs, or cubes and form again as Widmanstatten patterns. Some examples are shown in Fig. 42. There is no reason why the different stages of the ageing sequence should have sharp boundaries and two or more transition phases may frequently co-exist although such a state cannot, of course, continue indefinitely. An example of this situation is shown in Fig. 42(*a*) where small regions of G.P. zones are still visible in between the Widmanstatten γ' precipitates. It is noticeable that the G.P. zones nearest the γ' plates have already dissolved to give a 'denuded zone'.

Figure 42(*c*) shows a Widmanstatten precipitate of the η (MgZn$_2$) precipitate in two grains divided by a grain boundary and it is clear that there are no precipitates near the boundary. This example of a 'precipitate free zone' is commonly found in two-phase alloys and was first discussed in detail by Geisler.[118] One obvious way in which such a zone could form is by solute denudation caused by the large-grain boundary precipitates. Although this process occurs, particularly in over-aged alloys, many precipitate free zones are still supersaturated and the absence of

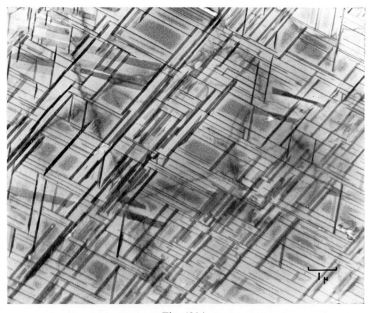

Fig. 42(*a*)

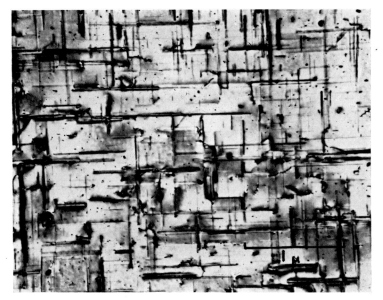

Fig 42(b)

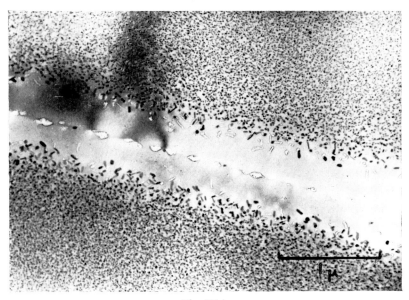

Fig. 42(c)

Fig. 42. Illustrating the formation of transition precipitates in age-hardening alloys: (a) γ' in Al–Ag (note the remaining G.P. zones in gaps between the γ' plates); (b) β' in Al–Mg–Si (Lorimer);[125] and (c) η' in Al–Zn–Mg (note the precipitate free zone at the grain boundary) (Lorimer).[125]

precipitates is due to the absence of suitable nucleating sites as first suggested by Rosenbaum and Turnbull.[119] The latter hypothesis can be used to explain the microstructure of Al–Zn–Mg alloys.[111,114] Local changes in the microstructure at grain boundaries often have deleterious effects on the properties of the alloy, for example its resistance to stress corrosion and fatigue.[120]

The equilibrium precipitate will normally have a structure which is completely different from the matrix. Hence all its interfaces are likely to be non-coherent and the particle shape is usually spherical or irregular.

4.2.1. *Heterogeneous nucleation*

Figure 42(c) showed that preferential nucleation of precipitates occurs on grain boundaries during the decomposition of supersaturated solid solutions. As shown in Section 2, the nucleation rate of the equilibrium phase on a crystal defect may well be larger than the homogeneous nucleation rate of a transition phase so that it is common to find extensive grain boundary precipitation of the equilibrium phase and a fine dispersion of a transition phase in the matrix (Fig. 43(a)). Grain boundary precipitates are often lens shaped with a constant dihedral angle where the particle boundaries and the grain boundary meet at a triple point. This indicates that the

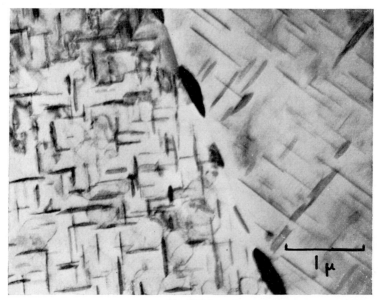

Fig. 43(a)

Fig. 43(b)

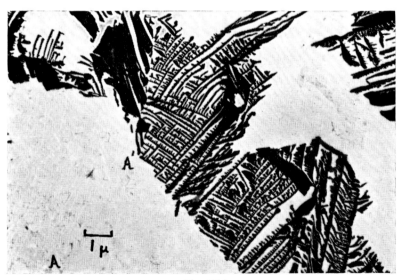

Fig. 43(c)

Fig. 43. Preferential precipitation of the equilibrium phase at grain boundaries: (a) θ particles in Al–Cu (the transition phase θ' has formed in the grains); (b) α particles in γ-iron in a 0·29%C 0·70%Mn steel (Aaronson);[122] and (c) dendritic precipitation of $Cr_{23}C_6$ in an unstablised stainless steel (Hatwell and Berghezan).[123]

particle has an equilibrium shape, as discussed on page 259, and observations of this type have been used to make relative measurements of interface energies.[121]

Grain boundary particles may often take shapes which are clearly not related to equilibrium but may be dictated by the atomic structure of the boundary, the crystallographic orientation of the adjacent grains or the kinetics of solute atom migration. Pro-eutectoid ferrite particles in austenite show these variations rather well,[122] *e.g.* Fig. 43(*b*), but the most spectacular arrays are found with alloy carbides in steels, Fig. 43(*c*). These variations are of considerable interest from the point of view of particle morphology, but they are also important in determining many of the properties of alloys such as the ductility, fracture toughness, hot workability and resistance to intergranular corrosion.

The other important heterogeneous nucleation site in alloys is the dislocation. Extensive precipitation on dislocations can occur at supersaturations where the matrix nucleation rate is negligible.[26] At one time this was used as the principal technique in the study of dislocations in crystals. Figure 44(*a*) shows a dislocation network in an AgBr crystal where the dislocations are visible by virtue of their decoration by precipitates of silver. Similar effects can occur in alloys particularly at low supersaturations and above the G.P. zone solvus (Fig. 44(*b*)). Both Figs. 44(*a*) and 44(*b*) show examples where the homogeneous nucleation rate of precipitates is negligible. Under other conditions precipitation can occur both homogeneously and heterogeneously with the nucleation frequency of the latter being higher. Equally the lower activation energy for nucleation may allow precipitation of the equilibrium phase or a late transition phase on dislocations while matrix precipitation is confined to G.P. zones. An example of this situation is shown in Fig. 44(*c*).

Precipitation on dislocations can lead to small volumes of the material having a coarse structure while the remainder shows a fine homogeneous dispersion of precipitates. This can lead to some loss of properties. Alternatively, the process may cause a much finer dispersion than is possible without dislocations because of the low rate of matrix nucleation. In these circumstances, plastic deformation prior to ageing will lead to a finer dispersion of precipitates after ageing and a consequent increase in properties. The best known example of the commercial use of this technique is the ausforming of steel,[126] where carbide precipitates are nucleated on dislocations introduced by plastic deformation of the metastable austenite.[127] Another example, where improved ambient and high-temperature properties result, is the repeated precipitation of carbides on

Fig. 44(a)

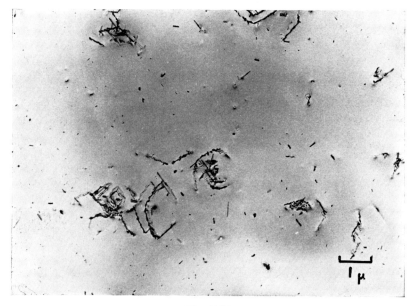

Fig. 44(b)

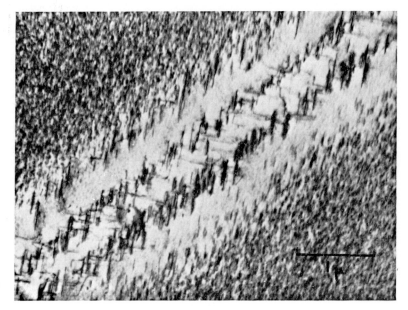

Fig. 44(*c*)

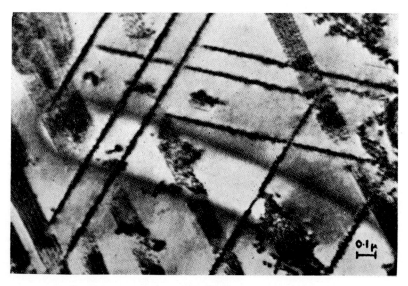

Fig. 44(*d*)

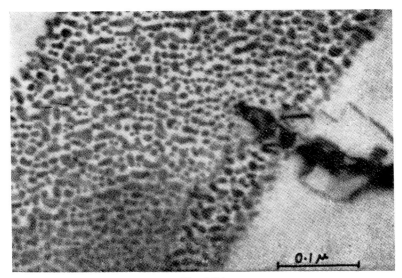

Fig. 44(e)

Fig. 44. Illustrating nucleation of precipitates on crystal dislocations (a) Ag particles in AgBr (Amelinckx).[124] (b) MgZn$_2$ particles in Al–Zn–Mg (Lorimer)[125] (c) θ' particles in Al–Cu (background is θ''). (d and e) Low magnification and high magnification of 'stacking faults precipitates' of NbC in an 18%Cr 10%Ni 1% Nb steel. (Van Aswegen, Honeycombe and Warrington).[130]

moving dislocations in austenitic stainless steels.[128–130] This microstructure (Fig. 44(d)) has been called 'stacking fault precipitation', a misleading name since the precipitates are nucleated on a moving Frank partial dislocation and the stacking fault is only formed by the enforced climb of this dislocation. The net result of the repeated nucleation of precipitates on the partial dislocation is the apparent formation of Widmanstatten plates (Fig. 44(d)), which in reality consist of a dense planar array of very small carbide particles (Fig. 44(e)).

4.2.2. Discontinuous precipitation

We noted earlier that precipitation may take place discontinuously at an interface which migrates through the material. The most common interface to find in this reaction is the grain boundary, since this exists in the material at the start of decomposition and, as discussed earlier, is a potent site for early nucleation of the equilibrium precipitate. The boundary can then migrate through the crystal leaving a mixture of the equilibrium phases behind it with a rod-shaped or lamellar morphology in a very similar manner to eutectic solidification (Section 3.3). The process is simpler than eutectic solidification in one sense because there is no possibility of convective mixing in the solid parent phase, but it is experimentally more complex owing to the difficulty of controlling the growth which takes place in random directions on a fine scale. Figure 45(a) is an example of the partial completion of a discontinuous reaction while Fig. 45(b) shows the interface between the transformed and untransformed material in more detail. The driving force comes from the reaction:

$$\alpha_{\text{supersat}} \rightarrow \alpha_{\text{sat}} + \beta \qquad (33)$$

This reaction is, of course, competitive with the continuous precipitation reaction given in eqn. (32). The main difference between these reactions is that continuous precipitation will normally occur *via* a series of transition phases whereas the equilibrium precipitate forms immediately in a discontinuous reaction. This difference in driving force is partly balanced by the large volume over which the continuous reaction can take place. Thus it is not easy to predict which reaction will dominate in different circumstances. A small amount of discontinuous precipitation is extremely common at the start of the decomposition process since the driving force for the reaction given by eqn. (33) will then be a maximum. As the supersaturation of the matrix is reduced by continuous precipitation, the driving force for the discontinuous reaction is also reduced and it is likely to cease after only a small volume of the alloy has decomposed.

The control of discontinuous precipitation is important in practice since it often leads to grain boundary embrittlement of an otherwise strong alloy. For example a small addition of Co is made to commercial Cu–Be alloys since it has been found to prevent the grain boundary reaction.[131]

4.2.3. *Spinodal decomposition*

Following the discussion of spinodal decomposition in Section 2.5, we will now consider the microstructure produced by this type of reaction. First of all, we will assume that the factor $E/(1 - v)$ in eqn. (20) is not a function of direction in the material, *i.e.* the material is elastically isotropic. An example of such a material is glass. Certain glasses have phase diagrams of the type shown in Fig. 9 and are found to decompose spinodally at sufficiently large supercooling. One might expect the microstructure to be rather complex since, as eqn. (20) shows, all wavelengths greater than λ_c are stable at a given supercooling and this spectrum of wavelengths must be present in all directions in the material. The situation is greatly simplified by the kinetics of the decomposition. Although the driving force is greatest for the largest wavelength, the diffusion distance is least for the smallest

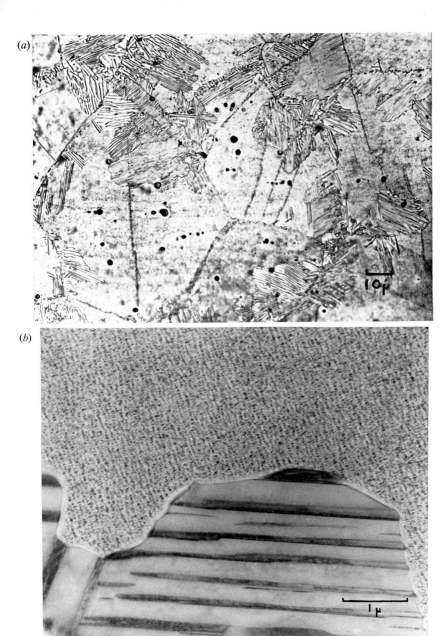

Fig. 45. Disontinuous precipitation in an Ni–Cr–Ti alloy: (a) optical micrograph showing partial transformation to a lamellar structure; and (b) electron micrograph showing the discontinuous Ni$_3$Ti lamellae consuming the continuous transition phase γ' (Merrick[150]).

stable wavelength. A comparison of these two factors shows that the fastest growing wavelength has a value $\sqrt{2}\lambda_c$[30] and this is the one observed in metallographic experiments, although the progress of the full spectrum can be followed by X-ray techniques.[134] The result of a large number of waves of wavelength $\sqrt{2}\lambda_c$ randomly superposed is a series of irregularly shaped islands of a second phase spaced approximately by $\sqrt{2}\lambda_c$ and this is the structure observed in spinodally decomposed glasses (Fig. 46(a)).

In crystalline materials, it is rare for $E/(1 - v)$ to be constant with direction in the crystal[135] and most metallic and non-metallic crystals are markedly anisotropic. As an example, let us consider the situation in cubic metals where $E/(1 - v)$ is nearly always a minimum in the $\langle 100 \rangle$ directions. This means that, for a given supersaturation, λ_c is a minimum in these directions and hence decomposition is most rapid for waves along $\langle 100 \rangle$. In practice, the anisotropy is so marked that waves in other directions can be neglected and the microstructure is built up by the superposition of three orthogonal waves of wavelength $\sqrt{2}\lambda_c$. Cahn[136] has shown that the distribution in space of second phase material which results from this decomposition is such that at small volume fractions the microstructure consists of an array of 'particles' arranged at the points of a simple cubic 'lattice' of edge length $\sqrt{2}\lambda_c$, while at larger volume fractions these particles link up along $\langle 100 \rangle$ to form an array of rods rather like builders' scaffolding. An example of this microstructure is shown in Fig. 46(b). Two other salient features of spinodal microstructures are noticeable from Fig. 46. First, the structure is remarkably uniform even compared with the fine-scale nucleation and growth structures such as are shown in Figs. 41 and 42. Secondly, there is no effect of lattice imperfections as is shown by the behaviour of the grain boundary in Fig. 46(b)—there is no change in the size and spacing of the precipitates only in their orientation. This characteristic is easily explained. Lattice imperfections assist the *nucleation* of a phase transformation and if this step is absent (as in spinodal decomposition) we would not expect them to have a significant effect on the phase change.

These characteristics of spinodal microstructures are likely to be significant in the development of certain properties, *e.g.* strength,[138] fatigue resistance[139] and magnetic hardness.[137] The potential of this type of phase change is just starting to be exploited.

4.2.4. *Eutectoidal decomposition*

The decomposition of a solid phase by a eutectoid reaction is the precise solid state analogue of eutectic solidification considered in Section 3.3.

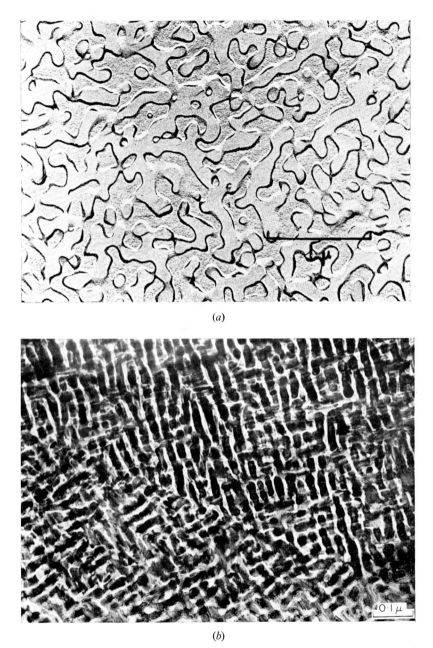

(a)

(b)

Fig. 46. Spinodal decomposition in (a) a non-crystalline glass (Cahn and Charles);[132] and (b) a crystalline Cu–Ni–Fe alloy (note the negligible effect of the grain boundary on the transformation).

It is also similar to discontinuous precipitation discussed earlier except that the decomposition reaction is now:

$$\gamma \to \alpha + \beta \tag{34}$$

which should be compared with eqn. (33). The moving interface now separates phases of different crystal structures instead of a common phase with different orientations, *i.e.* there is no longer a 'matrix' phase. However, this change has little effect on the nature of the interface and thus the kinetics of the reaction are similar to discontinuous precipitation.

Morphologically, eutectoid decomposition is very similar to eutectic solidification with the dispersed phase being lamellar or rod-like in shape (Fig. 47). The transformation normally nucleates heterogeneously at the grain boundaries of the high-temperature phase and then grows into one of the adjoining grains. The fineness of the structure is a function of the transformation temperature and typically the lamellar spacing d is related to the undercooling ΔT by[140]:

$$d \propto (\Delta T)^{-1} \tag{35}$$

More exactly the growth rate of the transformation must be taken into account and the presence of these three variables has led to a number of theories of the kinetics of eutectoid decomposition.[6]

One important difference between eutectoid decomposition and discontinuous precipitation is that the former cannot be prevented by quenching. Since the high temperature phase is unstable rather than merely supersaturated, the driving force for decomposition is so large that if a diffusional transformation is prevented by rapid quenching, a martensitic transformation takes place as described in Chapter 4. This is readily illustrated by quenching a sample after it has been partially transformed by a diffusional process when the two morphologies of transformation product resulting from the diffusional and martensitic changes are readily apparent.

We have so far considered the decomposition of an alloy of exactly eutectoid composition. If the composition is either side of eutectoid the high-temperature phase initially becomes unstable with respect to one of the low temperature phases only and this is formed essentially by a pre-cipitation reaction as described in Section 4.2. This pro-eutectoid reaction can, therefore, lead to grain boundary or Widmanstatten precipitation of, say, the α phase prior to transformation of the remaining γ to a eutectoid mixture of $\alpha + \beta$.

Eutectoid structures are most commonly found in steels where the rapid diffusion of the interstitial carbon atoms allows the decomposition to take

Fig. 47(a)

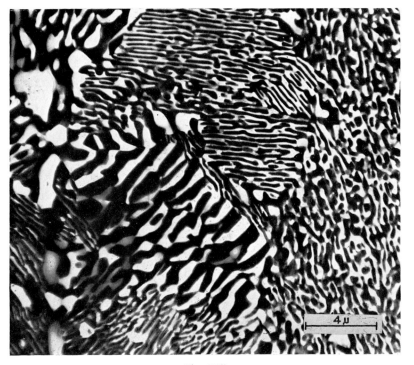

Fig. 47(b)

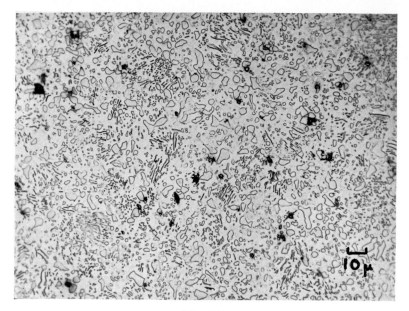

Fig. 47(c)

Fig. 47. Illustrating eutectoid decomposition in (a) 0·8%C steel to give pearlite (Vilella, Guellich and Bain),[146] and (b), 80%Zn–20%Al alloy (Nuttall)[153]. The lamellar structure of (a) may be spherodised to give the structure shown in (c).

place in reasonable times. In materials where the solvent is substitutionally dissolved, the massive redistribution of solute required by the transformation at a relatively low temperature makes the reaction so slow as to be uneconomic as a method of attaining properties and martensitic or fine-scale precipitation reactions are usually preferred.

4.3. Decomposition subsequent to a martensitic transformation

Although martensitic transformations are dealt with in detail in Chapter 4, it is worth mentioning here that the subsequent decomposition of a martensitic structure from its metastable state to the equilibrium mixture of two phases is identical in principle to the precipitation reactions considered in Section 4.2. There are two important differences in detail:

(a) The supersaturation, and hence the driving force, is usually much higher since it arises from the difference in solubility of a given element in the two allotropic phases rather than just the temperature variation of the solubility in a single phase.

(b) There is likely to be a high uniform distribution of lattice defects such as dislocations arising from the martensitic transformation and this makes it possible to obtain a fine precipitate dispersion by heterogeneous nucleation on these defects.

The net result of these differences is that the decomposition of a meta-stable martensite can often lead to an unusually fine distribution of a phase which is markedly dissimilar from the matrix to produce an alloy with excellent mechanical properties. This principle has been primarily applied to alloys based on iron and the tempering of quenched steel to produce a fine carbide dispersion in ferritic iron (Fig. 48(a)) is very well known. More recently the same principle has been applied to ferrous alloys with substitutional solutes to produce a dispersion of an intermetallic precipitate (Fig. 48(b)). This process has been called maraging[141] since the structure is produced by ageing the martensite—it results in some of the strongest commercial alloys available at the present time.

4.4. The analytical description of a solid–solid transformation

Although this chapter is primarily concerned with the *structure* resulting from phase changes in the solid state, it is pertinent to consider whether the

Fig. 48(a)

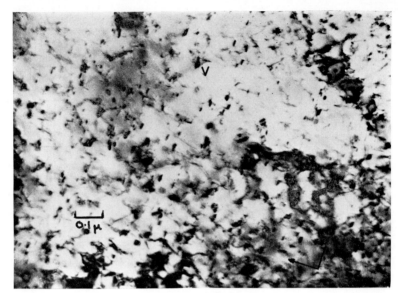

Fig. 48(b)

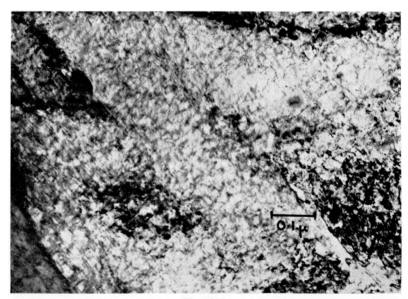

Fig. 48(c)

Fig. 48. Illustrating precipitation of (a) cementite, Fe_3C, in a quenched and tempered plain carbon steel (Tekin and Kelly);[151] (b) V_4C_3 in a secondary hardened steel (Tekin and Kelly);[151] and (c) intermetallics in a carbon free 18%Ni maraging steel (Miller and Mitchell).[152]

progress of the complex changes described in the preceding sections can be described by a single analytical expression. Avrami[142-144] and Johnson and Mehl[145] showed that a wide variety of solid state changes could be described by an equation:

$$\zeta = 1 - \exp{(-kt^n)} \tag{36}$$

where ζ is the fraction of the transformation completed in time t and k and n are constants which depend on G (eqns. (30) and (31)) and I (eqn. (16)). In fact the constant n depends primarily on the geometry of the transformed regions, *i.e.* whether, for example, the transformation is spreading from a number of points in the matrix in which case the transformed regions will be approximately spherical or whether it is nucleated on dislocations in which case the transformed regions will be cylindrical. For this reason the experimental determination of n is of some assistance in determining the average morphology of the transformation product.[6] It can readily be measured from a plot of ζ against t (Fig. 49(a)), by plotting $\log \log{(1/(1 - \zeta))}$ *versus* $\log t$ when the gradient is n (Fig. 49 (b)). The quantity ζ is most easily determined by following the change in some physical or mechanical property (*e.g.* resistivity or hardness) as a function of time during the reaction and equating $\zeta = 1$ with the total change in that property.

The Avrami analysis has obvious limitations since we have seen that the structure of an alloy is seldom homogeneous and several types of reactions may be competitive and proceed to different extents in different parts of the material or under different transformation conditions. Nevertheless, it remains the best general analytical expression to describe both the nucleation and growth stages of a solid–solid transformation.

4.5. Morphological changes subsequent to a phase transformation

Even when a phase transformation has resulted in equilibrium so far as the types of phases are concerned, the structure may still change with time as the morphology and distribution of the phases approaches equilibrium. We have seen that the morphology of the dispersed phase in a two-phase structure may be determined primarily by the kinetics and mode of transformation rather than by the thermodynamic requirement for minimum free energy. A good example is the lamellar morphology of a eutectoid reaction where the surface energy stored is greatly in excess of the amount which would be necessary if the dispersed phase were spherical in shape. This excess energy provides the driving force for a morphological change subsequent to the completion of the phase transformation. If the particle

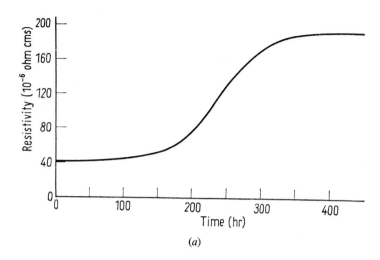

(a)

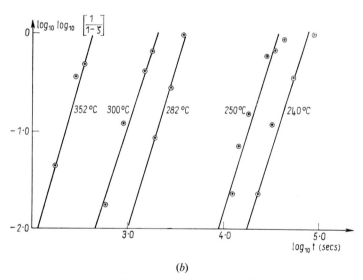

(b)

Fig. 49.   Showing (a) a typical result in following the progress of a solid state phase change as a function of time and (b) Avrami plots of data such as (a) (Christian).[6]

is non-coherent, the equilibrium shape is likely to be approximately spherical and hence this process is known as spherodisation (Fig. 47(c)).

If the particle has reached an equilibrium shape, *i.e.* the minimum surface energy per unit volume (Section 4.1), there is still excess energy stored by virtue of the degree of dispersion of second phase. Ideally, all the second

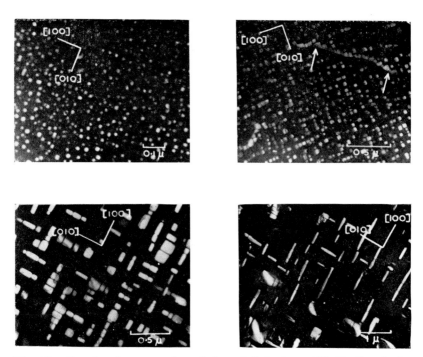

Fig. 50. Showing the coarsening of γ' precipitates in an Ni–Al alloy during ageing.

phase in the system should agglomerate into one large particle and although this is never found in practice, appreciable coarsening of a dispersed phase occurs at all temperatures where diffusion is possible (Fig. 50). The driving force for coarsening is the reduction in surface energy, and the mechanism by which it takes place depends on the small excess solute concentration required to maintain the equilibrium of a curved surface as opposed to a flat one (the Gibbs–Thomson effect). This results in concentration gradients being set up in the matrix which lead to diffusion currents and the progressive dissolution of small particles and the growth of large ones. The analysis of this situation has been carried to a fairly sophisticated level[147] and there is considerable quantitative evidence for the correctness of the theories of coarsening.[148,149]

Coarsening of a structure is extremely important in determining the variation of properties of a material with time since particle size and dispersion are always important parameters.

5. REFERENCES

1. Cottrell, A. H. (1955). *Theoretical Structural Metallurgy*, Arnold, London, p. 139 *et seq.*
2. Darken, L. S. and Gurry, R. W. (1953). *Physical Chemistry of Metals*, McGraw-Hill, New York, p. 326 *et seq.*
3. Volmer, M. and Weber, A. (1926). *Z. Phys. Chem.*, **119,** 227.
4. Becker, R. and Döring, W. (1935). *Ann. Phys.*, **24,** 719.
5. Hirth, J. P. and Pound, G. M. (1963). *Condensation and Evaporation*, Pergamon, London, p. 19 *et seq.*
6. Christian, J. W. (1965). *The Theory of Transformations in Metals and Alloys*, Pergamon, London, pp. 393 *et sq.*
7. Wulff, G. (1901). *Z. Krist*, **34,** 449; see also reference (6) pp. 146–52.
8. Turnbull, D. and Fisher, J. C. (1949). *J. Chem. Phys.*, **17,** 71.
9. Turnbull, D. (1950). *J. Appl. Phys.*, **21,** 1022.
10. Walker, J. W. (1966). See K. A. Jackson *et al.* (1966), *Trans. Met. Soc. AIME*, **236,** 149.
11. Duwez, P., Willens, R. H. and Klements, W., Jnr. (1960). *J. Appl. Phys.*, **31,** 1136.
12. Duwez, P. and Willens, R. H. (1963). *Trans. Met. Soc. AIME*, **227,** 362.
13. Willens, R. H. (1962). *J. Appl. Phys.*, **33,** 3269.
14. Kingery, W. D. (1960). *Introduction to Ceramics*, Wiley, London, p. 314.
15. Volmer, M. (1929). *Z. Electrochem.*, **35,** 555.
16. Turnbull, D. and Vonnegut, B. (1952). *Ind. Eng. Chem.*, **44,** 1292.
17. Sundquist, B. E. and Mondolfo, L. F. (1961). *Trans. Met. Soc. AIME*, **221,** 157.
18. Sundquist, B. E. and Mondolfo, L. F. (1961). *Trans. Met. Soc. AIME*, **221,** 607.
19. Mondolfo, L. F. (1965). *J. Aust. Inst. Met.*, **10,** 169.
20. Eshelby, J. D. (1957). *Proc. Roy. Soc.*, **A241,** 376.
21. Clemm, P. S. and Fisher, J. C. (1955). *Acta Met.*, **3,** 70.
22. Smith, C. S. (1953). *Trans. ASM*, **45,** 533.
23. Russell, K. C. (1969). *Acta Met.*, **17,** 1123.
24. Nicholson, R. B. (1970). *Phase Transformations*, p. 269, ASM, Cleveland.
25. Cottrell, A. H. (1948). *Strength of Solids*, Physical Society, London, p. 30.
26. Cahn, J. W. (1957). *Acta Met.*, **5,** 188.
27. Kelly, A. and Nicholson, R. B. (1963). *Prog. Mat. Sci.*, **10,** 151.
28. Beton, R. C. and Rollason, E. C. (1957–58). *J. Inst. Met.*, **86,** 77.
29. Cahn, J. W. (1962). *Acta Met.*, **10,** 907.
30. Cahn, J. W. (1968). *Trans. Met. Soc. AIME*, **242,** 166.
31. Cahn, J. W. and Hilliard, J. E. (1958). *J. Chem. Phys.*, **28,** 258; and (1959), **31,** 688.
32. Borelius, G., Andersson, J. and Gullbergy, K. (1943). *Ing. Vet. Akad. Hand.*, 169.
33. Jackson, K. A. (1958). *Liquid Metals and Solidification*, ASM, Cleveland, p. 174.
34. Burton, W. K., Cabrera, N. and Frank, F. C. (1951). *Phil. Trans. Roy. Soc.*, **A243,** 249.
35. Wagner, R. S. (1965). *Trans. Met. Soc. AIME*, **224,** 1182.

36. Jackson, K. A., Uhlmann, D. R. and Hunt, J. D. (1967). *J. Crystal Growth*, **1**, 1.
37. Miller, W. A. and Chadwick, G. A. (1969). *Proc. Roy. Soc.*, **A312**, 257.
38. Jackson, K. A. and Hunt, J. D. (1965). *Acta Met.*, **13**, 1212.
39. Cahn, J. W. (1960). *Acta Met.*, **8**, 554.
40. Cahn, J. W., Hillig, W. B. and Sears, G. W. (1964). *Acta Met.*, **12**, 1421.
41. Jackson, K. A. and Chalmers, B. (1956). *Can. J. Phys.*, **34**, 473.
42. Hillig, W. B. and Turnbull, D. (1956). *J. Chem. Phys.*, **24**, 219.
43. Volmer, M. I. and Mander, M. (1931). *Z. Physik. Chem.*, **A154**, 97.
44. Weinberg, F. and Chalmers, B. (1951). *Can. J. Phys.*, **29**, 382.
45. Jackson, K. A. (1962). *Phil. Mag.*, **7**, 1117.
46. Young, F. W. and Savage, J. R. (1964). *J. Appl. Phys.*, **35**, 1917.
47. Teghtsoonian, E. and Chalmers, B. (1951). *Can. J. Phys.*, **29**, 270.
48. Doherty, P. E. and Chalmers, B. (1962). *Trans. Met. Soc. AIME*, **224**, 1124.
49. Jaffrey, D. and Chadwick, G. A. (1968). *Phil. Mag.*, **18**, 573.
50. Tiller, W. A., Rutter, J. W., Jackson, K. A. and Chalmers, B. (1953). *Acta Met.*, **1**, 428.
51. Chalmers, B. (1964). *Principles of Solidification*, John Wiley, New York, Chapter 5.
52. Tiller, W. A. and Rutter, J. W. (1956). *Can. J. Phys.*, **34**, 96.
53. Davies, G. J. (1968). *The Solidification of Metals*, ISI Publication 110, p. 66.
54. Mullins, W. W. and Sekerka, R. F. (1964). *J. Appl Phys.*, **35**, 444.
55. Glicksman, M. E. and Vold, C. L. (1968). *The Solidification of Metals*, ISI Publication 110, p. 37.
56. Hellawell, A. and Herbert, P. M. (1962). *Proc. Roy Soc.*, **A269**, 560.
57. Chadwick, G. A. (1963). *Prog. Mat. Sci.*, **12**, 97.
58. Bell, J. A. E. and Winegard, W C. (1965). *J. Inst. Met.*, **93**, 457.
59. Rumball, W. M. and Kondic, V. (1966). *Trans. Met. Soc. AIME*, **236**, 586.
60. Hunt, J. D. and Jackson, K. A. (1966) *Trans. Met. Soc. AIME*, **236**, 843.
61. Hunt, J. D. and Chilton, J. P. (1962–63). *J. Inst. Met.*, **91**, 338.
62. Chadwick, G. A. (1962–63). *J. Inst. Metals*, **91**, 169.
63. Tiller, W. A. (1958). *Liquid Metals and Solidification*, ASM Cleveland, p. 276.
64. Jackson, K. A. and Hunt, J. D. (1966). *Trans. Met. Soc. AIME*, **236**, 1129.
65. Chadwick, G. A. (1963–64). *J. Inst. Met.*, **92**, 18.
66. Moore, A. and Elliott, R. (1968). *The Solidification of Metals*, ISI Publication 110, p. 167.
67. Day, M. G. and Hellawell, A. (1968). *Proc. Roy. Soc.*, **A305**, 473.
68. Day, M. G. (1969). *J. Metals*, **21**(4), 31.
69. Day, M. G. (1970). *J. Inst. Met.*, **98**, 57.
70. Plumb, R. C. and Lewis, J. E. (1957–58). *J. Inst. Met.*, **86**, 393.
71. Kim, C. B. and Heine, R. W. (1963–64) *J. Inst. Met.*, **92**, 367.
72. Cooksey, D. J. S., Munson, D., Wilkinson, M. P. and Hellawell, A. (1964). *Phil. Mag.*, **10**, 745.
73. Kerr, H. W., Plumtree, A. and Winegard, W. C. (1964–65). *J. Inst. Met.*, **93**, 63.
74. Cooksey, D. J. S. and Hellawell, A. (1967). *J. Inst. Met.*, **95**, 183.
75. Mollard, F. R. and Flemings, M. C. (1967). *Trans. Met. Soc. AIME*, **239**, 1526, 1534.
76. Collins, W. T., Jnr. and Mondolfo, L. F. (1965). *Trans. Met. Soc. AIME*, **233**, 1671.
77. Bates, H. E., Wald, F. and Weinstein, M. (1969). *J. Mat. Sci.*, **4**, 25.
78. Hellawell, A. (1967). *Met. and Materials*, **1**, 361.
79. Davies, G. J. (1971). *Strengthening Methods in Crystals* (Ed. by A. Kelly and R. B. Nicholson). Elsevier, London, Chapter 8.
80. Uhlmann, D. R. and Chadwick, G. A. (1961). *Acta Met.*, **9**, 835.
81. Sartell, J. A. and Mack, D. J. (1964–65). *J. Inst. Met.*, **93**, 19.
82. Delves, R. T. (1965). *Brit. J. Appl. Phys.*, **16**, 343.

83. Chadwick, G. A. (1965). *Brit. J. Appl. Phys.*, **16**, 1095.
84. Livingston, J. D. and Cline, H. E. (1969). *Trans. Met. Soc. AIME*, **245**, 351.
85. Chalmers, B., see reference (51), Chapter 8.
86. Chalmers, B. (1963). *J. Aust. Inst. Met.*, **8**, 255.
87. Jackson, K. A., Hunt, J. D., Uhlmann, D. R. and Seward, P. T., III (1966). *Trans. Met. Soc. AIME*, **236**, 149.
88. Cole, G. S. and Bolling, G. F. (1965). *Trans. Met. Soc. AIME*, **233**, 1568.
89. Flemings, M. C. (1968). *The Solidification of Metals*, ISI Publication 110, p. 277.
90. Cole, G. S. and Bolling, G. F. (1966). *Trans. Met. Soc. AIME*, **236**, 1366.
91. Kattamis, T. Z. and Williamson, R. B. (1968). *J. Inst. Met.*, **96**, 251.
92. Frawley, J. J. and Childs, W. J. (1968). *Trans. Met. Soc. AIME*, **242**, 256.
93. Chalmers, B. (1965). *Liquids: Structures, Properties, Solid Interactions* (Ed. by T. J. Hughel). Elsevier, p. 308.
94. Wojciechowski, S. and Chalmers, B. (1968). *Trans. Met. Soc. AIME*, **242**, 690.
95. Chadwick, G. A. (1969). *Met. and Materials*, **3**, 77.
96. Hall, H. T. and Jackson, W. J. (1968). *The Solidification of Metals*, ISI Publication 110, p. 313.
97. Frank, F. C. (1963). *The Relation between the Structure and Mechanical Properties of Metals*, NPL Symposium No. 15, HMSO, London, p. 248.
98. Turnbull, D. (1955). *Impurities and Imperfections*, ASM, Cleveland, p. 121.
99. Van der Merwe, J. H. (1963). *J. Appl. Phys.*, **34**, 117.
100. Becker, R. (1938). *Ann. der Physik*, **32**, 128.
101. Nicholson, R. B., *Interfaces*, Butterworths, Melbourne, 1970.
102. Laird, C. and Aaronson, H. I. *Acta Met.*, in press.
103. Laird, C. and Aaronson, H. I. (1968). *Trans. Met. Soc. AIME*, **242**, 1437.
104. von Schreiber, C. *Beyträge zur Geschichte und Kentniss meteorischer Stein- und Metall-masen, Vienna*, 1820.
105. Axon, H. J. (1968). *Prog. Mat. Sci.*, **13**, 185.
106. Barrett, C. S. and Massalski, T. B. (1966). *Structure of Metals* 3rd Ed., McGraw-Hill.
107. Rowcliffe, A. F. (1969). Ph.D. Thesis, University of Manchester.
108. Guinier, A. (1959). *Solid State Physics*, **9**, 294.
109. Nabarro, F. R. N. (1940). *Proc. Roy. Soc.*, **A175**, 519.
110. Nabarro, F. R. N. (1940). *Proc. Phys. Soc.*, **52**, 90.
111. Embury, J. D. and Nicholson, R. B. (1965). *Acta Met.*, **13**, 403.
112. Lorimer, G. W. and Nicholson, R. B. (1966). *Acta Met.*, **14**, 1009.
113. Lorimer, G. W. and Nicholson, R. B. (1969). *The Mechanism of Phase Transformations in Crystalline Solids*, Institute of Metals, London, p. 36.
114. Holl, H. A. (1964–65). *J. Inst. Metals*, **93**, 364.
115. Nicholson, R. B. and Nutting, J. (1961). *Acta Met.*, **9**, 332.
116. Nicholson, R. B. and Nutting, J. (1958). *Phil. Mag.*, **3**, 531.
117. Weatherly, G. C. (1966). Ph.D. Thesis, University of Cambridge.
118. Geisler, A. H. (1951). *Phase Transformations in Solids*, John Wiley, p. 387.
119. Rosenbaum, H. S. and Turnbull, D. (1959). *Acta Met.*, **7**, 664.
120. Thomas, G. and Nutting, J. (1959–60). *J. Inst. Metals*, **88**, 81.
121. Smith, C. S. (1952). *Imperfections in Nearly Perfect Crystals*, John Wiley, p. 377.
122. Aaronson, H. I. (1956). *The Mechanisms of Phase Transformations in Metals*, Institute of Metals, London, p. 47.
123. Hatwell, H. and Berghazan, A., *Precipitation Processes in Steels*, ISI Special Report No. 64, 1959, p. 88.
124. Amelinckx, S. (1958). *Acta Met.*, **6**, 34.
125. Lorimer, G. W. (1968). Ph.D. Thesis, University of Cambridge.
126. Shyne, J., Zackay, V. F. and Schmatz, D. (1960). *Trans. ASM*, **52**, 346.

127. Johari, O. and Thomas, G. (1965). *Trans. ASM*, **58**, 563.
128. Silcock, J. M. (1963). *JISI*, **201**, 409.
129. Pickering, F. B. (1963). *The Relation between the Structure and Mechanical Properties of Metals*, NPL Symposium No. 15, HMSO, London, p. 397.
130. Van Aswegen, J. S. T., Honeycombe, R. W. K. and Warrington, D. H. (1964). *Acta Met.*, **12**, 1.
131. Robertson, W. D. and Bray, R. S. (1959). *Precipitation from Solid Solution*, ASM Cleveland, p. 328.
132. Cahn, J. W. and Charles, R. J. (1965). *J. Phys. Chem. Glasses*, **6**, 181.
133. Tomozawa, M., Herman, H. and MacCrone, R. K. (1969). *The Mechanism of Phase Transformations in Crystalline Solids*, Institute of Metals, London, p. 6.
134. Rundman, K. B. and Hilliard, J. E. (1967). *Acta Met.*, **15**, 1025.
135. Schmid, E. and Boas, W. (1935). *Plasticity of Crystals*, Springer-Verlag, Berlin.
136. Cahn, J. W. (1961). *Acta Met.*, **9**, 795.
137. Nicholson, R. B. and Tufton, P. J. (1966). *Z. fur angew. Phys.*, **21**, 59.
138. Zackay, V. F. and Parker, E. R. (1965). *High Strength Materials*, John Wiley, p. 130.
139. Ham, R. K., Kirkaldy, J. S. and Plewes, J. T. (1967). *Acta Met.*, **15**, 861.
140. Zener, C. (1946). *Trans. AIME*, **167**, 550.
141. Floreen, S. (1968). *Met. Reviews*, **13**, 115.
142. Avrami, M. (1939). *J. Chem. Phys.*, **7**, 1103.
143. Avrami, M. (1940). *J. Chem. Phys.*, **8**, 212.
144. Avrami, M. (1941). *J. Chem. Phys.*, **9**, 177.
145. Johnson, W. A. and Mehl, R. F. (1939). *Trans. Met. Soc. AIME*, **135**, 416.
146. Vilella, J., Guellich, S. and Bain, E. C. (1936). *Trans. ASM*, **24**, 225.
147. Greenwood, G. W. (1969). *The Mechanism of Phase Transformations in Crystalline Solids*, Institute of Metals, London, p. 103.
148. Ardell, A. J. (1969). *The Mechanism of Phase Transformations in Crystalline Solids*, Institute of Metals, London, p. 111.
149. Ardell, A. J. and Nicholson, R. B. (1966). *Acta Met.*, **14**, 1295.
150. Merrick, H. F. (1963). Ph.D. Thesis, University of Cambridge.
151. Tekin, E. and Kelly, P. M. (1965). *Precipitation from Iron-Base Alloys*, Gordon and Breach, p. 173.
152. Miller, G. P. and Mitchell, W. I. (1965). *J. Iron Steel Inst.*, **203**, 899.
153. Nuttall, K. (1969). Ph.D. Thesis, University of Manchester.

CHAPTER 6

STRUCTURAL DEFECTS IN NONSTOICHIOMETRIC COMPOUNDS

B. T. M. WILLIS AND J. WILLIAMS

1. INTRODUCTION

The concept of a stoichiometric compound stems from one of the most important laws given in elementary inorganic chemistry text books, namely, the Law of Definite Proportions. This law had its roots in Dalton's atomic theory, and was only accepted after a long controversy between Proust and Berthollet. The issue was reopened in the early 1900s by Kurnakov, who discovered intermetallic compounds containing elements in nonstoichiometric proportions. In 1930 Schottky and Wagner showed, on the basis of classical statistical thermodynamics, that there is lattice* disorder in all crystals above $0°K$, so that crystalline inorganic compounds are intrinsically nonstoichiometric. Thus, the metallurgist and the solid-state chemist have approached the subject of nonstoichiometry from different directions: their common meeting ground has proved to be the defect structure of nonstoichiometric compounds.

For certain compounds, deviations from stoichiometry are very small and it is reasonable to assume that the defects are non-interacting and randomly distributed throughout the lattice. Examples are alkali halides containing F-centres and intrinsic semiconductors. For an ionic compound, MX, there are four possible types of point defects—M vacancies, X vacancies, interstitial M ions and interstitial X ions: the relative proportions of each can usually be determined from density measurements, diffusion studies, measurements of electrical conductivity, or from thermodynamic considerations. The characterisation of point defects in nonstoichiometric compounds in these ways has been reviewed by Libowitz.[1]

However, there are many substances showing *gross* deviations from ideal stoichiometry and for these the assumption of isolated point defects

* The term 'lattice' is used in a loose sense, where strictly we should write 'structure' (cf. Chapter 1).

is no longer valid. These materials include a large number of compounds of technological importance. Doped rare-earth and zirconium oxides are being exploited as furnace elements in the ducts of magnetohydrodynamic systems and in high-temperature fuel cells, where optimum properties are achieved in oxides having a very high concentration of defects. The actinide dioxides are the most important of the nuclear fuels: an unirradiated fuel such as UO_2 can exhibit gross departures from stoichiometry by taking up oxygen atoms into solution as UO_{2+x} (where x is as high as 0·25), and all fuels acquire a high concentration of lattice defects after long periods of irradiation. Very little is known of the basic defect structure of such materials. Even if the defects are regarded as randomly-distributed point defects, many of the defects must occupy adjacent lattice sites and so interact with one another producing some kind of 'defect complex'.

For the characterisation of the defects in grossly nonstoichiometric compounds, the diffraction method can be used, consisting of the measurement and interpretation of the number of X-ray quanta (or slow neutrons) scattered by the sample. The easiest approach is to determine the structure of a sample in which the defects have undergone long-range ordering. Thus ferrous sulphide is nonstoichiometric and is always iron-deficient ($Fe_{1-x}S$): Bertaut[2] showed that the cation vacancies are ordered in alternate layers of the ideal NaCl structure at the composition Fe_7S_8. Another way of ordering defects is to concentrate the vacancies into planes dividing the crystal into blocks of the ideal structure. These blocks are then sheared so as to eliminate the vacancies and to produce a 'shear structure' with a unit cell dependent on the block size. Shear structures were first found by Magneli[3] in the oxides of molybdenum and tungsten where a homologous series of different compositions occurs. The crystallographer naturally turns his attention to these superlattice or shear phases, as the determination of their structures can be undertaken by well-established procedures (*see* Chapter 1), applicable to perfect crystals. However, the long-range ordering *eliminates* the defects which we wish to characterise, and produces a perfect three-dimensional atomic arrangement. The relation between the nature of the structural defects and the structure of the highly-ordered phase in which they are absorbed is not necessarily a straight-forward one.

A much more difficult problem for the crystallographer is the characterisation of the structural defects in grossly nonstoichiometric compounds in which there is no long-range ordering of the defects. It is this problem which is discussed in the present chapter. Lipson and Lipson have reminded us in chapter 1 that: 'We have probably reached the end of an epoch . . . and (crystallographic) methods must now be extended to the more difficult problems of imperfectly crystalline matter'. Relatively little progress has been made in this field so far, and inevitably our discussion will be limited to the few systems, especially UO_{2+x} and $Fe_{1-x}O$, which have been examined. In the next section the basic diffraction theory will be covered in outline, and in Section 3 we show how this theory has been applied in practice to obtain structural information about the defects in grossly nonstoichiometric compounds.

2.   BASIC CONCEPTS OF DIFFRACTION THEORY

There are two fundamentally different approaches to the theory of diffraction by a crystalline solid. In the 'kinematic theory' it is assumed that the amplitude of scattering is so small that interaction between incident and scattered waves can be neglected. This is a good assumption for the majority of crystals examined by X-ray or neutron diffraction, although 'extinction corrections' to the diffraction intensities, arising from the breakdown of the kinematic theory, may be necessary for the strongest Bragg reflections. Even large crystals can be treated by the kinematic theory: faults, such as sub-grain boundaries or walls of dislocation lines, break up the crystal into small blocks about 1 μm in size and dynamic interaction between incident and scattered waves can occur only within each block. For the scattering of X-rays from large crystals (for example, silicon or germanium) in which the lattice is perfectly coherent over extended regions, the more complicated 'dynamic theory' must be used. Nonstoichiometric compounds, containing point defects or defect complexes, are intrinsically imperfect, and the available evidence indicates that they approximate closely to the 'ideally imperfect' crystals to which the kinematic theory properly applies.

In electron diffraction, the proportion of the radiation scattered by each atomic plane is much larger than for X-ray or neutron diffraction, and the dynamic theory provides the only satisfactory basis for interpreting the experimental intensities. It is for this reason that X-ray diffraction is far more widely used than electron diffraction in crystal structure determination, in spite of the advantages of electron diffraction in obtaining data rapidly and from small areas of sample.

We shall discuss only the kinematic theory and its application to the scattering of X-rays (or neutrons) from nonstoichiometric compounds. For further details of this theory the reader is referred to the standard book on diffraction theory by R. W. James.[4]

2.1.   General formula for scattered intensity from a single crystal

The structure factor F_m of the mth unit cell is proportional to the amplitude of the radiation scattered by that cell, taking a corner of the cell as origin. The scattered amplitude with respect to an arbitrary origin fixed in the crystal is

$$F_m \exp (i\mathbf{Q} \cdot \mathbf{r}_m) \tag{2.1}$$

where the term $\exp (i\mathbf{Q} \cdot \mathbf{r}_m)$ is a phase factor expressing the phase difference between the mth cell and the crystal origin. $\mathbf{Q}$ is defined by

$$\mathbf{Q} = \frac{2\pi}{\lambda} (\mathbf{k} - \mathbf{k}_0)$$

where $\mathbf{k}$, $\mathbf{k}_0$ are unit vectors in the directions of the diffracted and incident beams respectively, and λ is the wavelength. $\mathbf{r}_m$ is the vector from the crystal origin to the mth cell (*see* Fig. 1). If the cell contains only one atom, the

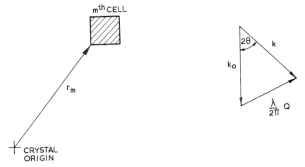

Fig. 1.   Definition of vectors in equation (2.1).

structure factor is equivalent to the atomic scattering factor f (X-rays) or the nuclear scattering amplitude b (neutrons).

The total scattered amplitude $Y(\mathbf{Q})$ is obtained by summing (2.1) over all cells

$$Y(\mathbf{Q}) = \sum_m F_m \exp (i\mathbf{Q} \cdot \mathbf{r}_m),$$

and the total scattered intensity, $I(\mathbf{Q})$, is $Y(\mathbf{Q}) Y^*(\mathbf{Q})$:

$$I(\mathbf{Q}) = \sum_m \sum_n F_m F_n^* \exp[i\mathbf{Q} \cdot (\mathbf{r}_m - \mathbf{r}_n)]$$

$$= \sum_m \sum_n F_m F_n^* \exp (i\mathbf{Q} \cdot \mathbf{R}_{mn}) \tag{2.2}$$

where $\mathbf{R}_{mn}$ is the vector joining cells m and n. Starred quantities in (2.2) represent complex conjugates.

2.2. Intensity from perfect crystal

For a three-dimensional crystal, consisting of rows of identical unit cells spaced at regular intervals apart, each cell has the same structure factor and so $F_m = F_n = F$. Neglecting thermal motion, the double summation in (2.2) works out as [5]

$$I(\mathbf{Q}) = |F|^2 \, \frac{\sin^2(\tfrac{1}{2}N_1\mathbf{Q}\cdot\mathbf{a}_1)}{\sin^2(\tfrac{1}{2}\mathbf{Q}\cdot\mathbf{a}_1)} \frac{\sin^2(\tfrac{1}{2}N_2\mathbf{Q}\cdot\mathbf{a}_2)}{\sin^2(\tfrac{1}{2}\mathbf{Q}\cdot\mathbf{a}_2)} \frac{\sin^2(\tfrac{1}{2}N_3\mathbf{Q}\cdot\mathbf{a}_3)}{\sin^2(\tfrac{1}{2}\mathbf{Q}\cdot\mathbf{a}_3)} \quad (2.3)$$

We have assumed, for convenience, that the crystal has the shape of a regular parallelepiped with edges $N_1\mathbf{a}_1$, $N_2\mathbf{a}_2$, $N_3\mathbf{a}_3$ where $\mathbf{a}_1$, $\mathbf{a}_2$, $\mathbf{a}_3$ are the edges of the unit cell. For N_1, N_2, $N_3 \gg 1$, there is constructive interference in the directions determined by the three Laue equations:

$$\left.\begin{array}{c} \mathbf{Q}\cdot\mathbf{a}_1 = 2\pi h \\[4pt] \mathbf{Q}\cdot\mathbf{a}_2 = 2\pi k \\[4pt] \mathbf{Q}\cdot\mathbf{a}_3 = 2\pi l \end{array}\right\} \quad (2.4)$$

and destructive interference in all other directions. The integers h, k, l in equations (2.4) are the indices of the Bragg reflecting planes, and it can be shown readily that these equations lead to the Bragg equation

$$\lambda = 2d \sin \theta \quad (2.5)$$

where θ is the glancing angle of incidence for radiation reflected by the planes of spacing d.

Thus, for an ideal crystal containing a regular array of identical unit cells, there is no scattering of the primary radiation unless the crystal is correctly oriented to satisfy Bragg's equation (2.5) for a particular family of planes (hkl). When this condition is satisfied, Bragg scattering occurs at an angle 2θ to the primary beam and in the plane defined by the primary beam and the (hkl) normal. As the wavelength is increased, fewer and fewer planes are available for Bragg scattering, until finally, for wavelengths exceeding the 'cut-off wavelength' λ_c, there is no possibility at all of scattering (*see* Fig. 2). From equation (2.5), λ_c is given by

$$\lambda_c = 2d_{\max} \quad (2.6)$$

where $d_{\max}$ is the maximum spacing of the reflecting planes in the crystal: typical values of λ_c lie in the range 4–7 Å. In X-ray diffraction, attenuation of the beam by true absorption (fluorescence) increases rapidly with wavelength and is much more important than attenuation by scattering: in neutron diffraction, absorption is usually extremely small, so that a neutron beam with a wavelength exceeding λ_c can pass with very little

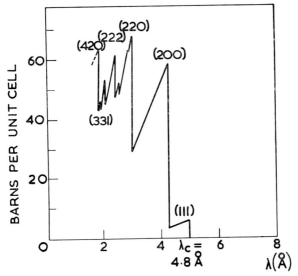

Fig. 2.   Calculated total neutron scattering cross-section for nickel oxide (rock salt structure) plotted versus wavelength. Sharp discontinuities occur where the wavelength just exceeds that corresponding to Bragg scattering at $\theta = 90°$ for the (hkl) planes shown. No Bragg scattering at all occurs for $\lambda > 4·84$ Å.

attenuation through a large perfect crystal. This property of neutron radiation is particularly important for the examination of defects in crystals by means of diffuse scattering (see below).

2.3.   Intensity from an imperfect, nonstoichiometric crystal

In equation (2.3), it is assumed that the structure factor is the same for all cells: this is not so for a crystal with defects, and so we must return to the general expression (2.2) to determine the scattered intensity. For a nonstoichiometric crystal, we shall describe the atomic arrangement in terms of an ideal structure, free from defects, which accommodates structural defects responsible for the deviation from stoichiometry. The structure factor of the mth cell can then be expressed as

$$F_m = \bar{F} + \Delta_m$$

where $\bar{F}$ is the structure factor averaged over all cells and Δ_m is the local deviation from the average caused by departure from the ideal atomic arrangement. Substituting this expression for the structure factor into equation (2.2):

$$I(\mathbf{Q}) = \sum_m \sum_n (\bar{F} + \Delta_m)(\bar{F}^* + \overline{\Delta_n^*}) \exp(i\mathbf{Q} \cdot \mathbf{R}_{mn}),$$

$$\text{or } I(\mathbf{Q}) = |\bar{F}(\mathbf{Q})|^2 \sum_m \sum_n \exp(i\mathbf{Q} \cdot \mathbf{R}_{mn})$$

$$+ \bar{F} \sum_m \sum_n \Delta_n^* \exp(i\mathbf{Q} \cdot \mathbf{R}_{mn})$$

$$+ \bar{F}^* \sum_m \sum_n \Delta_m \exp(i\mathbf{Q} \cdot \mathbf{R}_{mn})$$

$$+ \sum_m \sum_n \Delta_m \Delta_n^* \exp(i\mathbf{Q} \cdot \mathbf{R}_{mn}). \qquad (2.7)$$

The contributions of the four terms in (2.7) to the total scattered intensity $I(\mathbf{Q})$ will be considered separately.

2.3.1. *Bragg scattering: fundamental reflections*

The first term on the right-hand side of equation (2.7) is equivalent to the expression for the intensity scattered by a perfect crystal with a structure factor $\bar{F}(\mathbf{Q})$ for all cells. It reduces to expression (2.3) with $F = \bar{F}$, and is the only term in (2.7) which gives rise to sharp Bragg maxima. The interpretation of the Bragg intensities from a nonstoichiometric crystal, using the equation

$$I_{\text{Bragg}} = |\bar{F}(\mathbf{Q})|^2 \sum_m \sum_n \exp(i\mathbf{Q} \cdot \mathbf{R}_{mn}), \qquad (2.8)$$

will yield a set of structure factors $|\bar{F}_{hkl}|$ for each hkl plane. These structure factors are averaged over all cells in the crystal, so that the information given by the Bragg intensities relates to the 'average structure', which can be conceived as the structure representing the superposition of the contents of all the cells ($\sim 10^{23}$ in number) in the crystal. In a perfect crystal, all cells are identical and so the average structure is simply the contents of one cell. This simplification does not occur in a crystal with defects, and so it is not possible, *from the Bragg intensities alone,* to deduce the microscopic atomic arrangement.

2.3.2. Diffuse scattering

Turning to the remaining terms in (2.7), Δ_m and Δ_n are equally likely to be positive or negative, and so the second and third terms in (2.7) are separately zero. The last term, however, does not vanish and represents the contribution to the non-Bragg, diffuse scattering. It can be rewritten as follows:

$$I_{\text{diffuse}} = \sum_m \sum_n \Delta_m \Delta_n^* \exp (i\mathbf{Q} \cdot \mathbf{R}_{mn})$$

$$= \sum_m |\Delta_m|^2 + \underset{m \neq n}{\sum_m \sum_n} \Delta_m \Delta_n^* \exp (i\mathbf{Q} \cdot \mathbf{R}_{mn})$$

or

$$I_{\text{diffuse}} = N\langle |F_m - \bar{F}|^2\rangle + \underset{m \neq n}{\sum_m \sum_n} \Delta_m \Delta_n^* \exp (i\mathbf{Q} \cdot \mathbf{R}_{mn}) \qquad (2.9)$$

where the angle brackets $\langle\rangle$ denote an average over all N cells. If the deviation from stoichiometry is caused by isolated point defects, distributed at random throughout the crystal and producing no distortion of the lattice in their immediate neighbourhood, there is no correlation between the scattering from cells m and n and the quantity $\Delta_m \Delta_n^*$ averages out as nothing. In this case, equation (2.9) reduces to

$$I_{\text{diffuse}} = N\langle |F_m - \bar{F}|^2\rangle. \qquad (2.10)$$

(2.10) represents diffuse intensity uniformly distributed throughout reciprocal space. This intensity is known as the 'Laue Monotonic Intensity', and its effect is to raise the background level above which the Bragg reflections appear. However, relaxation of the lattice in the neighbourhood of a defect, or clustering of the defects into complexes, give rise to a correlation between Δ_m and Δ_n, especially when m and n refer to neighbouring cells, and so $\langle \Delta_m \Delta_n^*\rangle$ does not vanish. The net effect of lattice relaxation or of clustering is to cause modulations to the uniform Laue scattering (see Fig. 3). The only certain example of Laue monotonic scattering, unaccompanied by intensity modulations, is that arising in neutron diffraction where scattering occurs from an element containing isotopes of different nuclear scattering amplitudes. There is no reason to believe that the individual isotopes are not randomly distributed throughout the lattice, and so uniform Laue diffuse intensity results (Fig. 4). In nonstoichiometric crystals, lattice relaxation around the defects, or local defect clustering, will produce intensity modulations arising from the second term in (2.9).

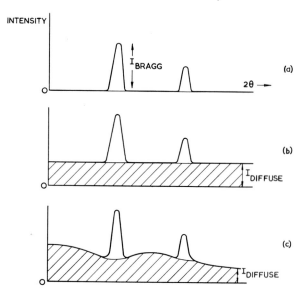

Fig. 3. Schematic (neutron) powder diffraction patterns of: (a) perfect crystal, showing no background between Bragg peaks; (b) imperfect crystal containing isolated point defects without lattice relaxation, showing uniform diffuse scattering; (c) imperfect crystal containing defect complexes, showing intensity-modulated diffuse scattering.

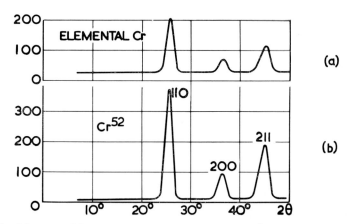

Fig. 4. Neutron diffraction powder pattern, $\lambda = 0\cdot9$ Å, from (a) elemental chromium, containing a mixture of the isotopes Cr^{50}, Cr^{52}, Cr^{53}, Cr^{54}; (b) the single isotope Cr^{52}. The high uniform background in (a) represents monotonic Laue diffuse scattering. (Courtesy of C. G. Shull).

To analyse these intensity effects, a plausible model describing the atomic rearrangement caused by the defects is postulated and the double sum in (2.9) evaluated. The procedure is similar to that used by Warren[5] in determining short-range order parameters from the X-ray diffuse scattering from alloys. In contrast to the situation with Bragg scattering, the analysis of diffuse scattering yields, in principle, direct information about the local atomic arrangement.

Unfortunately, the analysis is complicated by the presence of other components of diffuse scattering which are not directly associated with the defects. The most important of these is thermal diffuse scattering (TDS), which increases rapidly with rise of temperature and precludes the study of diffuse scattering on other than quenched samples. TDS arises from the interaction of the radiation with the elastic waves constituting the thermal energy of the crystal. The elastic waves produce a correlated motion of the atoms in different cells of the crystal, and thermal diffuse scattering occurs which is associated with the second term in equation (2.9).

We can summarise the results of this section by describing the various diffraction effects as follows.

1. *Perfect crystal*
Bragg scattering for $\lambda < \lambda_c$.
No Bragg scattering for $\lambda > \lambda_c$.
No diffuse scattering, apart from that arising from thermal agitation (*i.e.* thermal diffuse scattering or TDS).

2. *Imperfect crystal with defects*
Bragg scattering for $\lambda < \lambda_c$. The intensity of the Bragg peaks gives the contents of the 'average cell' but not the local atomic arrangement.
Diffuse scattering from the defects. The intensity of the diffuse scattering is related to the local atom–atom separations within the crystal.

3. CHARACTERISATION OF LATTICE DEFECTS BY THE DIFFRACTION METHOD

We shall now consider the analysis of the intensity effects described in section 2 and discuss the type of information which this analysis yields by referring to experimental work on particular systems.

3.1. Bragg scattering: fundamental reflections

The intensities of the fundamental Bragg reflections give the set of structure factors $|\bar{F}(\mathbf{Q})|$, or $|\bar{F}_{hkl}|$ (*see* equation (2.8)), and from these the scattering density averaged over all cells can be determined. Because of the relatively small influence of the lattice defects on the observed intensities, which are contributed mainly by the scattering from atoms in undisturbed lattice sites, intensity data of exceptionally high quality are required. If the deviation from stoichiometry is very small (for example, as in $TiO_{1.997}$), the diffraction method based on the measurement of Bragg intensities is insufficiently sensitive to characterise the nature of the lattice defects.

The mean structure factor $\bar{F}_{hkl}$, or the amplitude of scattering from the contents of the 'average cell', is given by

$$\bar{F}_{hkl} = \sum_j m_j f_j \exp 2\pi i (hx_j + ky_j + lz_j) T_j \qquad (3.1)$$

where $x_j\ y_j\ z_j$ are the co-ordinates expressed as fractions of the edges of the average cell. f_j is the scattering amplitude of the jth atom, and m_j its 'occupation number'. For a stoichiometric compound the unit cell contains an integral number of atoms, so that m_j is unity for an occupied site and zero otherwise; in nonstoichiometric compounds, there is the possibility of partial occupancy of atomic sites in the average cell, and this fractional occupancy is the 'occupation number'. T_j in equation (3.1) is the 'temperature factor' for the jth atom, accounting for the influence of thermal motion on the Bragg intensities.

The characterisation of the lattice defects requires the evaluation of the atomic co-ordinates $x_j\ y_j\ z_j$ and the site occupation numbers m_j. This can be done using either the Method of Least Squares or the Fourier Method.

3.1.1. *Method of least squares*

A least-squares comparison is made between the set of observed structure factors F_{obs} for the different families of reflecting planes hkl and the calculated structure factors based on equation (3.1). m_j, x_j, y_j, z_j and the scale factor s are adjusted so as to reduce the quantity

$$\sum_{\text{all } hkl\text{'s}} |sF_{obs} - |\bar{F}_{hkl}||^2 \qquad (3.1a)$$

to a minimum value. The scale factor is included in order to account for the fact that intensity measurements are usually made on a relative scale.

A special difficulty, which is associated with the term T_j in equation (3.1), occurs in applying the least squares method to nonstoichiometric compounds. It may be preferable to make the intensity measurements at high temperatures where there is a wider range of composition in the homogeneous solid solution. The correction factor T_j^{-1} increases rapidly with temperature and, although its precise magnitude may have little importance in determining the gross features of the structure, an incorrect value of T_j will lead to corresponding errors in m_j and x_j, y_j, z_j. In the customary procedure for interpreting diffraction data at room temperature, T_j is expressed as

$$T_j = \exp\left[-(b_{11}h^2 + b_{22}k^2 + b_{33}l^2 + b_{23}kl + b_{31}lh + b_{12}hk)\right] \quad (3.2)$$

where the six quantities b_{11}, b_{22} ... define the lengths and direction cosines of the three principal axes of the 'thermal vibration ellipsoid' of the atom. Equation (3.2) embodies the result that the effect of thermal agitation is to spread the scattering power of the atom over an ellipsoidal volume; however, the result is valid only within the limits of the 'harmonic approximation', which implies that the crystal has no thermal expansion and very high lattice conductivity. The harmonic approximation describes an idealised situation which is not even true at $0°K$ (because of zero-point motion) and departs progressively further from reality as the temperature rises. It is an adequate approximation for interpreting room-temperature intensity measurements of moderate precision, but cannot be used for the analysis of accurate high-temperature data. Anharmonic effects give extra terms in equation (3.2), which can be as large as the harmonic terms.[6]

3.1.2. Fourier method

Fourier inversion of the structure factors $\bar{F}_{hkl}$ yields directly the distribution of scattering density $\rho(xyz)$ in the average cell. Thus

$$\rho(xyz) = \frac{1}{V}\sum_h\sum_k\sum_l \bar{F}_{hkl} \exp\left[-2\pi i(hx + ky + lz)\right] \quad (3.3)$$

where V is the cell volume and xyz are the co-ordinates of any point in the average cell, expressed as fractions of the cell edges.

The experimental measurements consist of intensities, from which the amplitudes of the structure factors, $\bar{F}_{hkl}$, are derived but not (directly) their phases: this constitutes the notorious phase problem of crystallography (see Chapter 1). However, there is really no phase problem in studying the disordered structure of nonstoichiometric compounds, provided the crystal structure of the corresponding ideal atomic arrangement (without defects) is known. If we write

$$\bar{F}_{hkl} = F_{obs} \exp\left(i\alpha_{hkl}\right) \quad (3.4)$$

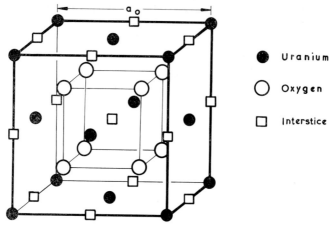

Fig. 5. Unit cell of UO_2, with the fluorite structure. The oxygen atoms are arrayed at the points of a simple cubic lattice of side $\frac{1}{2}a_0$, and the uranium atoms at the points of a face-centred lattice of side a_0. There are large interstices in the cell at the $\frac{1}{2}\frac{1}{2}\frac{1}{2}$ ... sites.

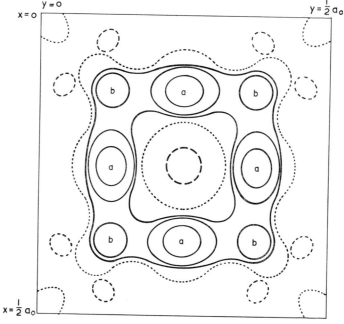

Fig. 6. Fourier section normal to [001] of nuclear density of UO_{2+x}, with contributions from uranium atoms and fluorite-type oxygen atoms removed. Peaks at a represent interstitial oxygen atoms O' and peaks at b interstitial oxygen atoms O''. Zero contours dotted, and negative contours broken.

of the interstitial oxygen atoms, which occupy *two* distinct types of crystallographic site, O′ and O″. The O′ sites are the positions

$$\tfrac{1}{2}vv,\ v\tfrac{1}{2}v,\ vv\tfrac{1}{2},\ \ldots \text{etc.}$$

of the $Fm3m$ space group, and the O″ sites the positions

$$www,\ w\overline{ww},\ \overline{w}w\overline{w},\ \ldots \text{etc.}$$

where $v \simeq w \simeq 0.4$.

Proceeding to the least squares comparison of observed and calculated structure factors (stage (ii) of Section 3.1.3), the results in Table 1 are obtained. The oxygen atom 0 is an anion in a normal fluorite-type site, and 0‴ corresponds to the largest interstice in the fluorite structure (indicated by □ in Fig. 5). The figures in brackets represent estimated standard deviations.

TABLE 1

$UO_{2.12}$: CONTENTS OF AVERAGE CELL

Atom	Co-ordinates in average cell			Contribution to oxygens in formula unit $UO_{2.12}$
0	$\tfrac{1}{4}$	$\tfrac{1}{4}$	$\tfrac{1}{4}$	1·87 (0·02)
0′	$\tfrac{1}{2}$	v	v	0·08 (0·04)
0″	w	w	w	0·16 (0·06)
0‴	$\tfrac{1}{2}$	$\tfrac{1}{2}$	$\tfrac{1}{2}$	0·00 (0·02)
	with			
	v = 0·38 (0·01)			
	w = 0·41 (0·01)			

The composition figures in the last column of Table 1, derived from the occupation numbers, show that there are three types of lattice defect—0 vacancies, 0′ interstitials and 0″ interstitials—and that the interstices at $\tfrac{1}{2}\tfrac{1}{2}\tfrac{1}{2}\ldots$ are unoccupied. Moreover, it is readily demonstrated that these defects must be associated into defect complexes in order to avoid the impossibly short interatomic distances arising from randomly distributed 0′ and 0″ interstial atoms.

The problem remains of determining the local arrangement of atoms in these defect complexes. We have seen that it is not possible to provide a unique answer from the analysis of the Bragg intensities alone, but plausible defect models can be postulated by combining the results in Table 1 with a knowledge of the ionic radii.

The structural model we shall discuss first is one in which the ratios of the concentrations, 0 vacancies: 0′ interstitials: 0″ interstitials, are rounded off to 2:1:2; the ratios as given in Table 1 are 13:8:16. The 2:1:2 ratios suggest that an interstitial atom enters the structure at an 0′ site, and thereby causes two fluorite-type oxygens to be ejected to two 0″ sites. This situation seems reasonable, as an 0′ atom can be comfortably placed in the $UO_{2.00}$ structure, provided the two nearest oxygens at a distance of 1·7 Å are removed (Fig. 7).

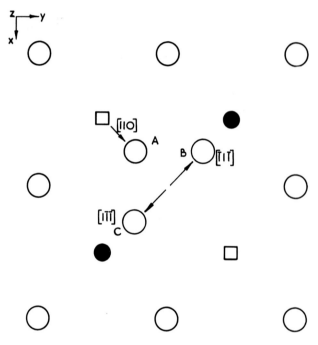

Fig. 7.   View of UO_2 structure along cubic axis. An interstital oxygen atom, O′, can be accommodated at A provided the two nearest oxygen atoms above and below (B and C) are displaced along ⟨111⟩ to 0″ sites.

A possible 2:1:2 structure based on this concept is shown in Fig. 8. The dotted lines in this diagram outline four cubes of side $\frac{1}{2}a_o$, where a_o is the cell size (5·47 Å), and these are labelled from the top left-hand corner as I, II, III and IV. In the $UO_{2.00}$ arrangement, oxygen atoms are located at the corners of each of the $\frac{1}{2}a_o$ cubes. In UO_{2+x} an extra oxygen atom enters the structure at the position marked E, where E is approximately 1 Å along the [110] direction from the centre of cube II. The two oxygens

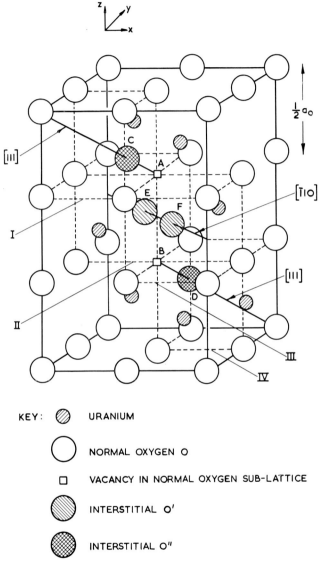

Fig. 8.   Defect complex in UO_{2+x} (after Roberts). [8]

at A and B are displaced along [111] to positions C, D on the body diagonals of cubes I, IV, respectively.

Using the numbers given in Table I, the precise atomic positions in the
2:1:2 defect complex can be specified, and the various atomic distances
evaluated. The minimum oxygen-uranium distance is 2·18 Å and the
minimum oxygen–oxygen distance is 2·24 Å: these are shorter than the
distances calculated from the ionic radii, but are much more reasonable
than the distances given by randomly distributed point defects.[8]

A difficulty with the 2:1:2 model is that, although the relative concentrations of defects are in approximate agreement with observation, the
absolute concentrations are predicted incorrectly. For $UO_{2\cdot12}$ the predicted
formula is $UO_{1\cdot76}O'_{0\cdot12}O''_{0\cdot24}$, whereas the observed formula is
$UO_{1\cdot87}O'_{0\cdot08}O''_{0\cdot16}$. The difficulty may be resolved as indicated in Fig. 8.
An extra O' atom can be inserted in cube III at the position F, where E
and F are equivalent sites related by 180° rotation about the line AB. This
leads to an alternative model for UO_{2+x}, in which the numbers of $0'$ and

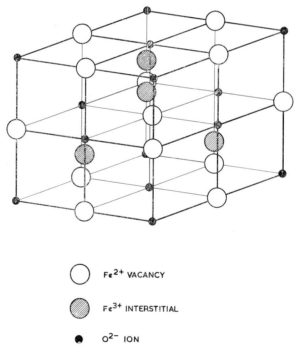

○ Fe^{2+} VACANCY

◍ Fe^{3+} INTERSTITIAL

● 0^{2-} ION

Fig. 9.   Cluster of thirteen vacancies and four interstitals in $Fe_{1-x}0$ (after
Koch and Cohen).[10]

O'' interstitials and O vacancies in the central defect complex are 2,2,2. For this $2:2:2$ model, the predicted formula is $UO_{1.88}O'_{0.12}O''_{0.12}$, which is in closer accord with the observed formula

$$UO_{1.87(0.03)}O'_{0.08(0.04)}O''_{0.16(0.06)}$$

There is some evidence that the type of defect complex discussed above is not restricted to the UO_{2+x} system. The same type occurs in solid solutions of YF_3 in CaF_2,[9] and it is possible that this complex is a general characteristic of the numerous binary and ternary compounds based on the fluorite structure with an excess of anions occupying interstitial positions.

In contrast to UO_{2+x}, the defects in wustite, $Fe_{1-x}O$, are restricted to the cation sublattice. However, the general conclusions from the X-ray diffraction study of $Fe_{0.90}O$ by Koch and Cohen[10] are similar to those from the neutron study of $UO_{2.12}$. The predominant proportion of defects in $Fe_{0.90}O$ occur as clusters rather than as isolated point defects, and each cluster contains 13 octahedrally-co-ordinated cation vacancies and 4 tetrahedrally-co-ordinated cation interstitials (Fig. 9). These clusters do not appear to be regions of magnetite, as suggested by the earlier neutron diffraction work of Roth[11] but bear a resemblance to them. The clusters persist at temperatures up to at least 1150°C.

3.2. Diffuse (non-Bragg) scattering

Here too there are two possible approaches to the analysis of the diffuse intensity: they are analogous to the least squares and Fourier methods used for the analysis of the Bragg intensity data. Either a structural model for the defects is postulated and the corresponding diffuse intensity is calculated and compared with the observed intensity, or the intensity is Fourier-inverted to yield the atomic distribution in the crystal. The Fourier method requires careful intensity measurements over a wide range of $Q(\equiv |Q| = 4\pi \sin \theta/\lambda)$ in order to avoid series-termination errors. It has been applied in one dimension to the interpretation of defect scattering from magnetic alloys[12] and from irradiated materials,[13] but the method is less generally used than that based on a trial-and-error determination of the defect model.

Childs[14] has undertaken an extensive study of the $Fe_{1-x}O$ system, using the long-wavelength neutron technique[13] with $\lambda > \lambda_c$. Because of the relative weakness of the diffuse scattering, it is important to avoid entirely the effect of Bragg reflections in measuring the defect scattering cross section. This is achieved by using 'cold' neutrons of long wavelength, λ.

Childs measured both the total cross section as a function of neutron wavelength and the differential cross section as a function of scattering angle for constant neutron wavelength. He examined samples in the composition range $Fe_{0.95}O$ to $Fe_{0.91}O$ and found that the observed cross sections could not be explained in terms of point defects. It was necessary to postulate clusters of defects, each similar to the Koch cluster illustrated in Fig. 9, but grouped together forming multi-clusters. The peak in the differential scattering cross-section at $Q = 0.55$ (Fig. 10) is due to short-

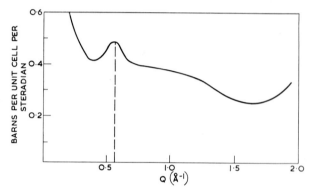

Fig. 10. Differential scattering cross section observed for $Fe_{0.92}O$ with 5·5 Å neutrons ($\lambda_c = 4.9$ Å) (after Childs[14]).

range ordering of the Koch clusters at a separation between the clusters of 17 Å.

A number of investigators have reported the observation of diffuse scattering arising from the structural defects in nonstoichiometric compounds, but the work of Childs represents the only detailed quantitative evaluation of the diffuse intensity. Many more diffraction studies are necessary, using the knowledge to be gained from analysing both the Bragg and diffuse intensity data, before general conclusions can be made regarding the nature of the structural defects. Only then will it be possible to make substantially further progress in understanding the thermodynamic and transport properties of grossly nonstoichiometric compounds.

4. REFERENCES

1. Libowitz, G. G. (1965). *Progr. Solid State Chem.*, **2**, 216.
2. Bertaut, E. F. (1953). *Acta Cryst.*, **6**, 557.
3. Magneli, A. (1949). *Arkiv Kemi*, **1**, 213.

4. James, R. W. (1962). *The Optical Principles of the Diffraction of X-rays*, Bell, London.
5. Warren, B. E. (1969). *X-ray Diffraction*, Addison-Wesley, New York.
6. Willis, B. T. M. (1969). *Acta Cryst. A.*, **25**, 277.
7. Willis, B. T. M. (1964). *Proc. Brit. Ceram. Soc.*, **1**, 9.
8. Roberts, L. E. J. (1965). *Thermodynamic and Transport Properties of Uranium Dioxide and Related Phases*, Chapter 2, IAEA, Vienna.
9. Cheetham, A. K., Fender, B. E. F., Steele, D., Taylor, R. I. and Willis, B. T. M. (1970). *Solid State Communications*, **8**, 171.
10. Koch, F. and Cohen, J. B. (1969). *Acta Cryst. B*, **25**, 275.
11. Roth, W. L. (1960). *Acta Cryst.*, **13**, 140.
12. Low, G. G. and Collins, M. F. (1963). *J. Appl. Phys.*, **34**, 1195.
13. Martin, D. G. (1964). *The Interaction of Radiation with Solids*, p. 643, North Holland, Amsterdam.
14. Childs, P. E. (1967). *Neutron Scattering by Nonstoichiometric Compounds*, Doctorate Thesis, University of Oxford.
15. Fender, B. E. F. and Henfrey, A. W. (1970). *J. Chem. Phys.*, **52**, 3250.

Note added in proof

The diffuse scattering studies of Dr. P. E. Childs have been extended further, to nonstoichiometric carbide and deuteride systems, by Drs. B. E. F. Fender and A. W. Henfrey.[15]

INDEX